L

Th
a

OVER-THE-COUNTER PHARMACEUTICAL FORMULATIONS

OVER–THE–COUNTER PHARMACEUTICAL FORMULATIONS

by

David B. Braun

np
NOYES PUBLICATIONS
Park Ridge, New Jersey, U.S.A.

Library of Congress Catalog Card Number: 94-2511
ISBN: 0-8155-1347-X
Printed in the United States

Published in the United States of America by
Noyes Publications
Mill Road, Park Ridge, New Jersey 07656

10 9 8 7 6 5 4 3 2 1

Library of Congress Cataloging-in-Publications Data

Braun, David B.
Over-the-counter pharmaceutical formulations / by David B. Braun.
p. cm.
Includes index.
ISBN 0-8155-1347-X
1. Drugs, Nonprescription. I. Title
RS125.B845 1994
615'.1373--dc20 94-2511
CIP

ABOUT THE AUTHOR

During his career as a research and development scientist, David B. Braun has worked in a broad spectrum of industries and technologies including rubber, plastics, pulp and papermaking, ceramics, water–soluble polymers and most recently, cosmetics and pharmaceuticals. He has written many technical papers for presentation to the various professional organizations associated with these industries and is the coauthor of a chapter on polyethers in the third edition of the *Kirk–Othmer Encyclopedia of Chemical Technology*. In addition, he has been awarded 10 United States and numerous foreign patents. He is also a member of the Society of Cosmetic Chemists and the Scientific Advisory Committee of the Cosmetic, Toiletry and Fragrance Association. He and his wife live in Connecticut.

To my dear wife, Lori
and our four admirable children,
Dianne, Katherine, Laura and David

PREFACE

This book presents formulations for over-the-counter (OTC) or non-prescription drugs. The phrase "over-the-counter" implies that the product may be purchased by the consumer without the need for a physician's prescription. The terms "nonprescription" and "OTC" are synonymous.

The OTC drug formulations in this book are classified according to their therapeutic effect. There are 19 categories of OTC drugs included. Each category is presented in a single chapter that consists of two parts; Part I presents the composition of brand name products, and Part II presents starting or prototype formulations contributed by suppliers of raw materials for OTC drugs.

This book includes the composition of 559 brand name OTC drugs produced by 64 manufacturers. There are also 270 suggested formulations contributed by 41 raw material suppliers. There is a total of 829 OTC pharmaceutical formulations.

The brand name products are listed alphabetically in Part I of each chapter, followed by the name of the manufacturer, the type and concentration of the active ingredient(s) and the product form(s). It is also customary, but not required, for the manufacturer to list the so-called "inactive ingredients" on the label of the product.

Part II of each chapter includes starting formulations, developed by a number of raw material suppliers. Not all chapters contain Part II because supplier's starting formulas for some OCT drug categories were not provided. Those formulations that are included, contain not only the concentration of the active ingredients but also the concentration of the other ingredients and, in most cases, a recommended procedure for mixing the formulation.

NOTICE

To the best of the Author's and Publisher's knowledge the information contained in this book is accurate; however, the Author and Publisher assume no responsibility nor liability for errors or any consequences arising from the use of the information contained herein. Final determination of the suitability of any information, procedure, or product for use contemplated by any user, and the manner of that use, is the sole responsibility of the user. The book is intended for informational purposes only. Due caution should be exercised in the use and handling of those raw materials that are potentially hazardous. Expert advice should be obtained at all times when manufacturing implementation is being considered. In the case of personal use of any of the products included, the manufacturer's medical instructions should be followed. Mention of trade names does not indicate endorsement by the Author or the Publisher.

Existing products are continually being improved with formulation changes. New products are being introduced and others are being withdrawn from the market. The reader is strongly advised to verify the accuracy of the composition of brand name products, or supplier's starting formulations presented by contacting the manufacturer or supplier, or by other means.

The suppliers of the starting formulations contained herein believe that the information they have provided is accurate and reliable. In most cases these formulations have been tested solely for rheology and stability, not for efficacy and safety. The suppliers do not warranty the results to be obtained with their products or formulas nor is freedom from infringement of any U.S. or foreign patent to be inferred. Since use conditions and local governmental regulations may vary from country to country and with time, it is the reader's responsibility to determine if the active ingredients and formulations included herein conform to current, local regulations.

CONTENTS

CHAPTER 1

INTRODUCTION

There is no question that the pharmaceutical industry provides an invaluable service to mankind. Without the products manufactured by this industry, there would be immeasurably greater pain and suffering than already exists in the world today. This book deals with the products and formulations of a major segment of this industry, Over-The-Counter (OTC) or nonprescription drugs.

We begin with the author's definition of OTC drugs which has guided the selection of the material contained herein. The phrase "over-the-counter" implies that the product may be purchased by the consumer without the need for a physician's prescription. The terms "nonprescription" and "OTC" are synonymous. The drugs described are developed by the manufacturer to favorably alter an existing, undesirable condition in the human body or to prevent or inhibit development of an undesirable human condition. A few examples will help to clarify this definition: internal analgesics such a Aspirin or Acetaminophen are usually ingested by the user to reduce the pain of headache or fever *after* the unpleasant condition develops. Dimenhydrinate is usually taken by the patient *in anticipation of* exposure to a situation that is known to cause nausea. No one would normally use a treatment for athlete's foot *before* the condition developed. Antiperspirants are applied *in anticipation of* excessive perspiration. Sunscreens are applied *before and during* sun exposure to prevent a burn from UV radiation.

An Industry Profile

This multi-billion dollar industry is made up of five distinct types of manufacturers. First, there are the large, multi-national corporations that are primarily engaged in the manufacturer of both prescription and OTC drugs, for example, Burroughs Wellcome, Inc. and Ortho Pharmaceutical Corp. Second, are the huge multi-national corporations that are involved in many different markets including the OTC drug market, such as Procter & Gamble Co. and Rhône-Poulenc, Inc. Third, are the smaller OTC drug companies that are primarily dedicated to that market, such as Whitehall Laboratories. Fourth, are the companies that are primarily cosmetic and toiletry manufacturers who also market some OTC products like sunscreens, antiperspirants, (classified as OTC drugs by the United States Food And Drug Administration) external analgesics and topical anti-infectives. These include Chesebrough-Ponds, Neutrogena Corp. and the like. Last, but by no means least, are the contract packagers that produce "generic" equivalents of the brand name products.

This book includes the composition of 559 brand name OTC drugs produced by 63 manufacturers that fall into the first four categories listed above. The composition of generic OTC drugs have not been included herein because it would be unnecessesarily repetitious. Virtually every drug store chain has its own version of many brand name products. There are also 270 starting formulations (contributed by 41 raw material suppliers) for a total of 829 OTC pharmaceutical formulations.

In the United States, there are four professional organizations that represent the interests of the OTC drug industry: The American Pharmaceutical Association (APhA) whose members include pharmacists, physicians and other interested professionals, The Pharmaceutical Manufacturers Association (PMA) whose 82 member companies (as of 8/92) include all the major producers of prescription and/or OTC drugs and the Nonprescription Drug Manufacturers Association (NDMA) whose 75 member companies (as of 9/92) include manufacturers of both brand name and/or generic OTC drugs. A fourth organization, The Cosmetic Toiletry And Fragrance Association (CTFA), with its approximately 240 members, also represents the OTC drug industry for those products that are usually manufactured by cosmetic companies, i.e. antiperspirants or sunscreens. All four associations are located in Washington, DC in close proximity to the legislative and regulatory powers that can profoundly affect the operations of the OTC drug industry.

Similar industry-oriented associations exist in the world's other industrialized nations.

Fifty years ago, OTC drugs were relatively few and available only at the local pharmacy. There was BAYER Aspirin, TUMS, PHILLIP'S Milk of Magnesia, to name a few. The advice of the resident pharmacist was frequently sought to guide the consumer's product selection. Today, OTC drugs are still available at the local pharmacy and the pharmacist is still there to assist the consumer in selection, but they may also be purchased at super markets, department stores, small "convenience" stores, gasoline stations and hotel gift shops. At these locations, there is no assistance from a pharmacist.

The number of OTC products has also proliferated over the past 50 years. As noted above, this book contains nearly 500 brand name products and there are many times this number of generic products. One reason is because the brand name product manufactures develop product line extensions by:

Varying the concentration of the active ingredient(s), i.e. regular and maximum strength.

Changing the combination of active ingredients, i.e cough and/or sinus and/or allergy versions.

Changing the product form, i.e. tablets and gelcaps.

To emphasize this point, the reader will find, near the end of Chapter 10, eight variations of TYLENOL cold and allergy treatment products (not counting variations in product form) and near the end Chapter 15 another six TYLENOL internal analgesic and antipyretic products. In addition, there is a continuing program of reclassification of prescription drugs to OTC status by the United States Food And Drug Administration. This offers an opportunity for the manufacturer to introduce entirely new OTC drug lines to the marketplace. This is discussed in greater detail later in this chapter.

The OTC Drug Review

The Food And Drug Administration (FDA), an agency of the United States Government, is responsible for regulation of the OTC drug industry (as well as the ethical drug industry). In 1972, the FDA initiated a program to review the active ingredients in OTC drugs to insure that they contain ingredients that are safe and effective and that the labeling on their packaging is fully informative. This program has come to be known as the OTC Drug Review.

This review is a three-phase rulemaking process culminating in the classification of active ingredients into one of three categories:

I. Generally recognized as safe and effective for the claimed therapeutic indication.

II. Not generally recognized as safe and effective or unacceptable indications.

III. Insufficient data available to permit final classification.

In the first phase of this review, a panel of experts reviews the available safety and efficacy data for a given active ingredient and publishes an Advance Notice of Proposed Rulemaking (ANPR) in the U.S. *Federal Register*. Comments from interested parties are solicited.

The second phase of the review consists of the consideration, by the FDA, of the comments received in response to the ANPR. Following this, the FDA prepares and publishes a Tentative Final Monograph (TFM) in the *Federal Register* and again invites comments from interested parties. New data may be submitted during this period as well as requests for a hearing before the FDA Commissioner.

In the final phase of the review, the FDA considers the comments received on the TFM and any new data submitted by interested parties and prepares and publishes a Final Monograph (Rule) in the *Federal Register* and eventually in the Code Of Federal Regulations (CFR).

The term "Final" for this monograph is a bit misleading because it is still possible for an OTC drug manufacturer or other interested party to petition the FDA to amend the Final Monograph based on new and pertinent data. Alternately, a manufacturer can circumvent the monograph by submitting a New Drug Application (NDA) for OTC use.

The Final Monograph usually contains four sections. The General Information section contains information about manufacturing practices, suitable inactive ingredients, color additives and similar regulations common to all OTC drugs. The Active Ingredients section specifies those active ingredients that are classified as Category I, generally recognized as safe and effective. The Labeling section contains the indications for the product, warnings against misuse, drug interactions and dosage frequency for different age groups. The monograph may also contain a fourth section containing specific testing methods that are required for the finished dosage form.

Occasionally, the OTC Drug Review process results in the conclusion that an ingredient should not be classified as Category I. The available data may not be sufficient to so classify the ingredient or the data may indicate that the ingredient is, in fact, *not* generally recognized as safe and effective. This results in the publication by FDA of a "Non-Monograph". Basically, this serves notice to the manufacturers that the use of the ingredient(s) and label claims of effectiveness and safety for the specified ingredient(s) will constitute misbranding of the product which could result in regulatory action by the FDA against the manufacturer.

Reclassification From Prescription To OTC Status

Since 1951, when federal laws more clearly defined the classes of drugs that were restricted to prescription use, the status of a number of prescription drugs has been changed, by the FDA, to OTC status. Federal regulations provide three routes to achieve prescription exemptions. The first is the so-called "switch" regulation that permits the FDA to exempt the drug from prescription-only use if the agency determines that the prescription requirement is unnecessary for the protection of the public health. The second route from prescription to OTC status is the filing, by the manufacturer, of an NDA for nonprescription use. The third route, which is being used to reclassify a number of drugs from prescription to OTC status, is through review and recommendation by the OTC Drug Review's Expert Advisory Panels. These ingredients are in various stages of the monograph development process described earlier. The reader is referred to the *Federal Register*, CFR and FDA Freedom of Information Office in Washington, DC for more detailed information on the current status of any ingredient in the OTC Drug Review process.

Format Of This Formulary

The OTC drug formulations in this book are classified according to their therapeutic effect, i.e. antidiarrheal products, cold and allergy products, etc. There are 19 categories of OTC drugs included herein. Each category is presented in a single chapter that consists of two parts; Part I presents the composition of brand name products and Part II presents starting or prototype formulations contributed by suppliers of raw materials for OTC drugs.

The brand name products are listed alphabetically in Part I of each chapter, followed by the name of the manufacturer, the type and concentration of the active ingredient(s) and the product form(s). It is also customary, but not required, for the manufacturer to list the so-called "inactive ingredients" on the label of the product.

It is the author's opinion that the term "inactive" is a misnomer. They should be referred to as "other" ingredients because the manufacturers do not normally add ingredients that perform no function to a formula. These other ingredients include colors, flavors, emulsifiers, suspending agents, binders and the like. Each of these ingredients usually provides some valuable function that is important to form, stability and/or consumer acceptance of the product. We have, therefore, chosen to list these as Other Ingredients. The nomenclature used in the Other Ingredients section of the formulation generally conforms to the C.T.F.A and /or U.S Pharmacopeia/ National Formulary nomenclaure systems. The former system is now referred to as International Nomenclature Cosmetic Ingredient (INCI).

Two non-standard abbreviations are included in Part I; "S.R." indicating a sustained release product and "Ch." indicating a "chewable" tablet. The other ingredients in the formula are usually, but not always, listed in alphabetical order, *not in order of decreasing concentration.* The concentration of the other ingredients in a brand name product is not public knowledge. The manufacturer usually provides this information to the FDA, but it can be accessed only with the written permission of the manufacturer.

Some brand name OTC drugs have not been included in this formulary for one or more reasons, i.e. the labeling does not reveal the concentration of the active ingredients. An exception to this has been made for the antiperspirants and sunscreens which do not normally indicate the concentration of the actives. The reader can, at the very least, determine the permissible concentration limits for the active ingredient(s) by consulting the pertinent FDA monograph on antiperspirants or sunscreens. Another reason for exclusion is that the product is not readily available at large retail outlets. This usually indicates that the product has been, or will shortly be, discontinued or replaced by an new and/or improved version.

Most suppliers of raw materials for the OTC drug industry maintain large, sophisticated research organizations to develop new raw materials and find new ways to use existing raw materials. As a part of this research, it is common for these supplier's researchers to develop starting or prototype formulas for testing new raw materials and/or to demonstrate to potential customers how a raw material can and should be used.

A number of raw material suppliers have generously contributed their starting formulations to this volume. These are presented in Part II of the appropriate chapter. All, except Chapter 17, contain a selection of the starting formulations. They contain not only the concentration of the active ingredients but also the concentration of the other ingredients and, in most cases, a recommended procedure for mixing the formulation.

In Part II, the trade name of the supplier's ingredients appear in bold face. Some supplier's formulations contain active ingredients manufactured by companies other than the supplier of the formulation. Since it is important that the formulator use the precise concentration of active ingredient in the formulation, we have included the trade name of the active ingredient used in parenthesis following the chemical name of the ingredient.

Appendixes

At the end of this volume, there are two appendixes. The first provides the reader with a list of all brand name OTC drugs manufacturers and their addresses. A second appendix provides a list of raw material suppliers that have contributed formulations to this book, with telephone and FAX numbers to contact. An Index of manufacturers, suppliers, brand name OTC drugs and active ingredients follows the appendixes.

FORMULATIONS

CHAPTER 2

ACNE TREATMENT PRODUCTS

The most frequently used active ingredients in OTC acne treatment products are Benzoyl Peroxide, Sulfur and Salicylic Acid. Although Resorcinol is claimed as an active in some products, "It is not generally regarded as efficacious in treatment of acne"[1]. Combinations of these ingredients are also available but, " the efficacy of these combinations over the single ingredient products has not been clearly demonstrated"[1]

Available product forms include creams, lotions, gels, ointments, cleansing bars and pads.

PART I

BRAND NAME FORMULATIONS

ACNOMEL

Manufacturer: Menley & James Laboratories, Inc.

Active Ingredients & Concentration:
- 8% Sulfur
- 2% Resorcinol

Product Form: Cream

Other Ingredients:
Alcohol (11% w/w), Bentonite, Fragrance, Iron Oxides, Potassium Hydroxide, Propylene Glycol, Titanium Dioxide, Purified Water.

ALMAY TROUBLE SPOTTER

Manufacturer: Almay, Inc.

Active Ingredient & Concentration:
- 3% Sulfur

Product Form: Cream

Other Ingredients:
Water, Cyclomethicone, Butylene Glycol, Talc, Stearyl Alcohol, Cetyl Alcohol, Kaolin, Magnesium Aluminum Silicate, Steareth-20, Allantoin, Steareth-2, Glyceryl Stearate, Xanthan Gum, Methylparaben, Ethylparaben, Imidazolidinyl Urea, Iron Oxides, Titanium Dioxide.

AMBI 10

Manufacturer: Kiwi Brands, Inc.

Active Ingredient & Concentration: 10% Benzoyl Peroxide

Product Form: Cream

Other Ingredients:
Aluminum Oxide, Bentonite, Carbomer 940, Glyceryl Stearate SE, Isopropyl Myristate, Methylparaben, PEG-12, Potassium Hydroxide, Propylparaben, Purified Water.

AVEENO, CLEANSING BAR FOR ACNE PRONE SKIN

Manufacturer: Rydelle Laboratories

Active Ingredient & Concentration: 2% Salicylic Acid

Product Form: Bar

Other Ingredients:
51% Colloidal Oatmeal, Sodium Cocoyl Isethionate, Purified Water, Glycerin, Lactic Acid, Sodium Lactate, Potassium Sorbate, Magnesium Aluminum Silicate, Titanium Dioxide, PEG-14M.

CLEAR BY DESIGN

Manufacturer: SmithKline Beecham Consumer Brands

Active Ingredient & Concentration: 2.5% Benzoyl Peroxide

Product Form: Gel

Other Ingredients:
Purified Water, Carbomer 940, Dioctyl Sodium Sulfosuccinate, Sodium Hydroxide, Edetate Disodium.

CLEARASIL, DOUBLE TEXTURED PADS, REGULAR STRENGTH

Manufacturer: Procter & Gamble Company

Active Ingredient & Concentration: 1.25% Salicylic Acid

Product Form: Pad

Other Ingredients:
Alcohol (40%), Water, Witch Hazel Distillate, Sodium Methyl Cocoyl Taurate, Quaternium-22, Aloe Vera Gel, Fragrance, Menthol, Edetate Disodium.

CLEARASIL, DOUBLE TEXTURED PADS, MAXIMUM STRENGTH

Manufacturer: Procter & Gamble Company

Active Ingredient & Concentration: 2.0% Salicylic Acid

Product Form: Pad

Other Ingredients:
Alcohol (40%), Water, Witch Hazel Distillate, Sodium Methyl Cocoyl Taurate, Quaternium-22, Aloe Vera Gel, Fragrance, Menthol, Edetate Disodium.

CLEARASIL, CLEARSTICK

Manufacturer: Procter & Gamble Company

Active Ingredient & Concentration: 1.25% Salicylic Acid

Product Form: Lotion

Other Ingredients:
Alcohol (40%), Water, Witch Hazel Distillate, Quaternium-22, Triethanolamine, Aloe Vera Gel, Fragrance, Menthol, Edetate Disodium.

CLEARASIL, CLEARSTICK MAXIMUM STRENGTH, Regular & Sensitive Skin

Manufacturer: Procter & Gamble Company

Active Ingredient & Concentration: Both Products: 2.0% Salicylic Acid

Product Form: Lotion

Other Ingredients:
Regular skin product contains Alcohol (39%), Water, Aloe Vera Gel, Triethanolamine, Fragrance, Menthol, Edetate Disodium.
Sensitive skin product also contains Glycerin but no Fragrance.

CLEARASIL, MAXIMUM STRENGTH, TINTED & VANISHING

Manufacturer: Procter & Gamble Company

Active Ingredient & Concentration	Product Form:
10% Benzoyl Peroxide	Cream

Other Ingredients:

Vanishing product contains Water, Propylene Glycol, Bentonite, Aluminum Hydroxide, Glyceryl Stearate SE, PEG 12M, Potassium Carbomer, Isopropyl Myristate, Methylparaben, Propylparaben.

Tinted Product also contains Dimethicone, Titanium Oxide, Iron Oxides.

EXACT

Manufacturer: Advanced Polymer Systems

Active Ingredient & Concentration	Product Form:
5.0% Benzoyl Peroxide	Cream

Other Ingredients:

Water, Acrylates Copolymer, Glycerin, Sorbitol, Cetyl Alcohol, Glyceryl Dilaurate, Stearyl Alcohol, Sodium Lauryl Sulfate, Magnesium Aluminum Silicate, Sodium Citrate, Silicon Dioxide, Citric Acid, Methylparaben, Xanthan Gum, Propylparaben.

FOSTEX (Benzoyl Peroxide Formula)

Manufacturer: Bristol-Myers Products

Active Ingredient & Concentration:	Product Forms:
10% Benzoyl Peroxide	Bar Gel

Other Ingredients:

Bar contains Boric Acid, Cellulose Gum, Dextrin, (may also contain wheat starch) Edetate Disodium, Dioctyl Sodium Sulfosuccinate, Lactic Acid, PEG-14M, Sodium Lauryl Sulfoacetate, or Sodium Dodecylbenzene Sulfonate, Disodium Sulfosuccinate, Sorbitol, Urea. Water.

Gel contains Carbomer 940, Diisopropanolamine, Edetate Disodium, Laureth-4, Water.

FOSTEX (Salicylic Acid Formula)

Manufacturer: Bristol-Myers Products

Active Ingredient & Concentration:	Product Form:
2.0% Salicylic Acid	Bar

Other Ingredients:
Boric Acid, Carboxymethylcellulose Sodium, Dextrin, Dioctyl Sodium Sulfosuccinate, Edetate Disodium, Fragrance, Sodium Lauryl Sulfoacetate, Octoxynol-2 Ethane Sulfonate, Sorbitol, Urea, Water.

LISTREX, HERBAL SCRUB

Manufacturer: Warner-Lambert Company

Active Ingredient & Concentration:	Product Form:
2.0% Salicylic Acid	Liquid

Other Ingredients:
1,3 Butylene Glycol, Colloidal Silicon Dioxide, FD&C Blue No. 1, FD&C Yellow No. 5, Fragrance, Methylparaben, Pentetete Pentasodium, TEA-Dodecylbenzenesulfonate, Water, Xanthan Gum, Triethanolamine.

NEUTROGENA, OIL-FREE ACNE WASH

Manufacturer: Neutrogena Corp.

Active Ingredient & Concentration:	Product Form:
2.0% Salicylic Acid	Lotion

Other Ingredients:
Purified Water, Sodium C14-16 Olefin Sulfonate, Cocamidopropyl Betaine, Sodium C12-15 Pareth-15 Sulfonate, Aloe Extract, Chamomile Extract, Matricaria Extract, Linoleylpropyl PG Dimonium Chloride Phosphate, Propylene Glycol, Edetate Disodium, FD&C Yellow No. 5, FD&C Red No. 40, Sodium Chloride, Fragrance.

NOXEMA, 2 IN 1 PADS, REGULAR STRENGTH

Manufacturer: Procter & Gamble Company

Active Ingredient & Concentration:	Product Form:
0.5% Salicylic Acid	Pad

Other Ingredients:
Alcohol (63%), Water, PPG-11 Stearyl Ether, Camphor, Clove Oil, Eucalyptus Oil, Menthol, Fragrance.

NOXEMA, 2 IN 1 PADS, MAXIMUM STRENGTH

Manufacturer: Procter & Gamble Company

Active Ingredient & Concentration:	Product Form:
2.0% Salicylic Acid	Pad

Other Ingredients:
Alcohol (63%), Water, PEG-4, PPG-11 Stearyl Ether, Camphor, Clove Oil, Eucalyptus Oil, Menthol, Fragrance.

OXY, MEDICATED CLEANSER

Manufacturer: SmithKline Beecham Consumer Brands

Active Ingredient & Concentration:	Product Form:
0.5% Salicylic Acid	Lotion

Other Ingredients:
Alcohol(40%), Citric Acid, Menthol, Propylene Glycol, Sodium Lauryl Sulfate, Water.

OXY, MEDICATED PADS, REGULAR STRENGTH

Manufacturer: SmithKline Beecham Consumer Brands

Active Ingredient & Concentration:	Product Form:
0.5% Salicylic Acid	Pad

Other Ingredients:
Alcohol (40%), Citric Acid, Fragrance, Menthol, Propylene Glycol, Sodium Lauryl Sulfate, Water.

OXY, MEDICATED PADS, MAXIMUM STRENGTH

Manufacturer: SmithKline Beecham Consumer Brands

Active Ingredient & Concentration:	Product Form:
2.0% Salicylic Acid	Pad

Other Ingredients:
Alcohol (50%), Citric Acid, Fragrance, Menthol, PEG-8, Propylene Glycol, Sodium Lauryl Sulfate, Water.

OXY, MEDICATED PADS, SENSITIVE SKIN

Manufacturer: SmithKline Beecham Consumer Brands

Active Ingredient & Concentration:	Product Form:
0.5% Salicylic Acid	Pad

Other Ingredients:
Alcohol (22%), Disodium Lauryl Sulfosuccinate, Fragrance, Menthol, PEG-4, Sodium Lauroyl Sarcosinate, Sodium PCA, Edetate Trisodium, Water.

OXY-5, TINTED & VANISHING

Manufacturer: SmithKline Beecham Consumer Brands

Active Ingredient & Concentration:	Product Form:
5% Benzoyl Peroxide	Cream

Other Ingredients:
Tinted product contains Cetyl Alcohol, Citric Acid, Iron Oxides, Methylparaben, Propylene Glycol, Propylparaben, Silica, Sodium Lauryl Sulfate, Stearyl Alcohol, Sodium PCA, Titanium Dioxide, Water, Vanishing product contains same ingredients as tinted product, but no Iron Oxides, Stearyl Alcohol or Titanium Dioxide.

OXY-10, TINTED & VANISHING

Manufacturer: SmithKline Beecham Consumer Brands

Active Ingredient & Concentration: 10% Benzoyl Peroxide

Product Form: Cream

Other Ingredients:
Tinted product contains Cetyl Alcohol, Citric Acid, Glyceryl Stearate, Iron Oxides, Methylparaben, Propylene Glycol, Propylparaben, Silica, Sodium Citrate, Sodium Lauryl Sulfate, Stearic Acid, Titanium Dioxide, Water,
Vanishing Product contains Carbomer 940, Dioctyl Sodium Sulfosuccinate, Edetate Disodium, Sodium Hydroxide, Water.

OXY-10, DAILY FACE WASH

Manufacturer: SmithKline Beecham Consumer Brands

Active Ingredient & Concentration: 10% Benzoyl Peroxide

Product Form: Lotion

Other Ingredients:
Citric Acid, Cocamidopropyl Betaine, Diazolidinyl Urea, Methylparaben, Propylparaben, Sodium Citrate, Sodium Cocoyl Isethionate, Sodium Lauroyl Sarcosinate, Water, Xanthan Gum.

OXY, NIGHT WATCH

Manufacturer: SmithKline Beecham Consumer brands

Active Ingredient & Concentration: 2.0% Salicylic Acid

Product Form: Lotion

Other Ingredients:
Cetyl Alcohol, Edetate Disodium, Methylparaben, Propylene Glycol, Propylparaben, Silica, Sodium Lauryl Sulfate, Stearyl Alcohol, Water.

STRIDEX, DUAL TEXTURED PADS, REGULAR STRENGTH

Manufacturer: Sterling Winthrop, Inc.

Active Ingredient & Concentration: 0.5% Salicylic Acid

Product Form: Pad

Other Ingredients:
Alcohol (28.25%), Citric Acid, Fragrance, Menthol, Purified Water, Simethicone Emulsion, Sodium Carbonate, Sodium Dodecylbenzenesulfonate, Sodium Xylenesulfonate.

STRIDEX, TEXTURED PADS, MAXIMUM STRENGTH

Manufacturer: Sterling Winthrop, Inc.

Active Ingredient & Concentration: 2.0% Salicylic Acid

Product Form: Pad

Other Ingredients:
Alcohol(44% w/w), Ammonium Xylenesulfonate, Citric Acid, Fragrance, Menthol, Purified Water, Simethicone Emulsion, Sodium Carbonate, Sodium Dodecylbenzenesulfonate.

STRIDEX, DUAL TEXTURED PADS, SENSITIVE SKIN

Manufacturer: Sterling Winthrop, Inc.

Active Ingredient & Concentration: 0.5% Salicylic Acid

Product Form: Pad

Other Ingredients:
Alcohol (28.25%), Aloe Vera Gel, Citric Acid, Fragrance, Menthol, Purified Water, Simethicone Emulsion, Sodium Carbonate, Sodium Dodecylbenzenesulfonate, Sodium Xylenesulfonate.

STRIDEX, SUPER SCRUB PADS, OIL FIGHTING

Manufacturer: Sterling Winthrop, Inc.

Active Ingredient & Concentration:	Product Form:
2.0% Salicylic Acid	Pad

Other Ingredients:
Alcohol (53.8%), Purified Water, Ammonium Xylenesulfonate, Citric Acid, Sodium Dodecylbenzenesulfonate, Sodium Lauroyl Sarcosinate, Sodium Carbonate, Menthol, Fragrance, Simethicone Emulsion.

PART II

SUPPLIER'S FORMULATIONS

ANTIACNE CREAM

Supplier: Amerchol Corp.

	Ingredient	Wt.%
A	Cetearyl Alcohol(and)Ceteareth 20 (**Promulgen D**)	5.00
	Diisopropyl Dimer Dilinoleate	1.50
	Water	67.70
B	Microcrystalline Cellulose(and)Cellulose Gum	2.80
C	Benzoyl Peroxide, 77%	13.00
D	Polysorbate 80(and)Cetyl Acetate (and) Acetylated Lanolin Alcohol (**Solulan 98**)	10.00
		100.00

Mixing Procedure:
Prepare Part C and Part D ahead of time using sufficient extra material to make up for mechanical losses in grind preparation. Grind Part C thoroughly in Part D. Pass through colloid mill to obtain desired particle size. Heat with stirring to 75°C. Cool to 60°C. Add Part B with good agitation until completely dispersed. Stir Parts C+D, grind to uniformity; add to A+B at 40°C. Cool to 30-35°C. Pass through a homogenizer.

ACNE CREAM

Supplier: American Colloid Company

	Ingredient	Wt.%
	Water	46.95
A	Carbomer 940	0.50
	Hydroxypropyl Methylcellulose, Medium Visc.	1.00
	Water	13.75
	Purified Bentonite (**Polargel NF**)	7.00
	Methylparaben	0.20
	Propylparaben	0.10
	Glyceryl Stearate	2.00
B	Propylene Glycol	6.00
	Polyethylene Glycol 600	2.00
	Myristyl Propionate	2.00
	Dimethicone	0.50
	Purified Bentonite	7.00
C	Titanium Dioxide	1.00
	Benzoyl Peroxide, 70%	10.00
		100.00

Mixing Procedure:

Sift Carbomer 940 into vortex of the water. When completely dispersed, sift in the Hydroxypropyl Methylcellulose. Add parabens to the water with stirring. Heat until dissolved with mixing. Add remaining Part B ingredients in the order listed above. With the addition of the **Polargel NF**, allow 15 minutes of mixing to complete hydration. Blend Part A with Part B. Add Benzoyl Peroxide and Titanium Dioxide. Mill.

ACNE LOTION

Supplier: American Colloid Company

	Ingredient	Wt.%
	Purified Bentonite (**Polargel NF**)	4.00
A	Hydroxypropyl Methylcellulose	1.00
	Water	52.20
	Water	19.00
	Methylparaben	0.20
	Propylparaben	0.10
B	Glyceryl Stearate	2.00
	Propylene Glycol	6.00
	Propylene Glycol 600	2.00
	Myristyl Propionate	2.00
	Dimethicone	0.50
	Iron Oxides	q.s.
C	Titanium Dioxide	1.00
	Benzoyl Peroxide, 77%	10.00
		100.00

Mixing Procedure:

Sift the **Polargel NF** into water with rapid mixing. Allow it to hydrate for 15 minutes water. Sift in the Hydroxypropyl Methylcellulose. Mix until completely smooth and there are no lumps. Add Parabens to the water with stirring. Heat with mixing until dissolved. Add remaining Part B ingredients in the order listed above. Blend Part A with Part B. Add Benzoyl Peroxide, Titanium Dioxide and Iron Oxides. Mill.

ACNE SCRUB CREAM

Supplier: American Colloid Company

	Ingredient	Wt.%
A	Magnesium Aluminum Silicate (**Magnabrite HV**)	2.00
	Water	58.20
B	Propylene Glycol	10.00
	Mineral Oil(and)Acetylated Lanolin Alcohol	15.00
	Glyceryl Stearate(and)PEG-100 Stearate	3.00
	Myristyl Propionate	1.40
C	Polyethylene	10.00
D	Eucalyptus Oil	0.40
	Preservative	q.s.
		100.00

Mixing Procedure:

Sift **Magnabrite HV** into the water slowly, agitating until smooth. Heat to 75°C. Heat Part B ingredients to 70°C. Add Part B to Part A with mixing until smooth. Add Part C and D to Parts A + B with mixing until cool.

ANTI-ACNE GEL

Supplier: American Distilling & Manufacturing, Inc.

	Ingredient	Wt.%
A	Distilled Witch Hazel, 14% Alcohol	42.40
	Salicylic Acid	0.50
	Aloe Vera Gel	5.00
	Sorbitol	1.00
B	Polyglycerylmethacrylate	50.00
C	Propylene Glycol	1.00
	Methylparaben	0.08
	Propylparaben	0.02
		100.00

Mixing Procedure:

Premix Part A ingredients. Add Part B with low shear, mixing until homogeneous. Premix Part C, and mix into the formulation. Package

---

KERATOLYTIC CREAM

Supplier: Croda, Inc.

	Ingredient	Wt.%
A	Emulsifying Wax NF (**Polawax**)	15.00
	PPG-2 Myristyl Ether Propionate (**Crodamol PMP**)	15.00
	Sorbitan Isostearate (**Crill 6**)	5.00
	Safflower Oil (**Super Refined Safflower Oil**)	3.50
	Avocado Oil (**Super Refined Avocado Oil**)	2.00
	Cetyl Palmitate (**Crodamol CP**)	2.00
	Salicylic Acid	5.00
	Propylparaben	0.15
	BHA	0.10
B	Deionized Water	48.75
	Sodium Borate	1.00
	Methylparaben	0.30
	Imidazolidinyl Urea	0.20
C	Hydrolyzed Collagen(and)Hyaluronic Acid (**Cromoist HYA**)	2.00
		100.00

Mixing Procedure:

Dissolve Salicylic Acid of Part A in **Crodamol PMP** with mixing and heat to 70°C. Add remaining ingredients of Part A with mixing and heat to 80°C. Combine ingredients of Part B with mixing and heat to 80°C. Add Part B to Part A with mixing and cool to 40°C. Add Part C with mixing and cool to desired fill temperature. Adjust pH, if necessary, to 3.0-4.0 with 10% TEA solution.

ANTI-ACNE CREAM

Supplier: Gattefossé s.a.

	Ingredient:	Wt.%
A	PEG-6(and)PEG-32(and)Glyceryl Stearate (**Tefose 63**)	20.00
	Mineral Oil	3.00
	Apricot Kernel Oil PEG-6 Esters (**Labrifil M 1944 CS**)	3.00
	Sorbic Acid	0.05
	Sodium Methylparaben	0.05
	Demineralized Water	72.40
B	Benzoyl Peroxide	0.50
	Econazole Nitrate	1.00
		100.00

Mixing Procedure:

Mix and heat the Part A ingredients to 75°C. Allow to cool while stirring. Add the Part B ingredients at 30°C and mix until uniform.

ANTI-ACNE LOTION

Supplier: Gattefossé s.a.

	Ingredient:	Wt.%
A	PEG-6 Stearate(and)Ceteth-20(and) Steareth-20 (**Tefose 2000**)	5.00
	Mineral Oil	3.00
	Cetyl Alcohol	2.00
B	Sodium Methylparaben	0.07
	Sorbic Acid	0.03
	Demineralized Water	88.40
C	Econazole Nitrate	1.00
	Benzoyl Peroxide	0.50
		100.00

Mixing Procedure:

Heat Part A and Part B separately to 75°C. Slowly add Part B to Part A while stirring. Cool with gentle stirring. Add Part C ingredients at 30°C while stirring.

ANTI-ACNE MICROEMULSION I

Supplier: Gattefossé s.a.

	Ingredient:	Wt.%
	Ethoxydiglycol (**Transcutol**)	47.00
A	PEG-8 Caprylic/Capric Glycerides (**Labrasol**)	25.00
	Propylene Glycol Dipelargonate (**D.P.P.G.**)	15.00
B	Benzoyl Peroxide	8.00
C	Propylene Glycol Laurate (**Lauroglycol**)	5.00
		100.00

Mixing Procedure:

Mix the Part A ingredients. Dissolve Part B in the Part A mixture with mixing (1.5 to 2 hours). Add Part C with mixing until uniform.

ANTI-ACNE MICROEMULSION II

Supplier: Gattefossé s.a.

	Ingredient:	Wt.%
	Ethoxydiglycol (**Transcutol**)	46.46
A	PEG-8 Caprylic/Capric Glycerides (**Labrasol**)	27.88
	Propylene Glycol Laurate (**Lauroglycol**)	4.64
B	Benzoyl Peroxide	7.44
C	Propylene Glycol Dipelargonate (**D.P.P.G.**)	9.30
	Propylene Glycol	4.28
		100.00

Mixing Procedure:

Mix the Part A ingredients. Dissolve Part B in the Part A mixture with mixing (2 hours). Add Part C ingredients with mixing until uniform.

ACNE CREAM

Supplier: Hüls America, Inc.

	Ingredient	Wt.%
A	Glyceryl Stearate SE (**Imwitor 960**)	10.00
	Propylene Glycol Dicaprylate/Dicaprate (**Myglyol 840**)	8.00
	Cetearyl Alcohol(and)Sodium Lauryl Sulfate	5.00
B	Propylene Glycol	3.00
	Allantoin	0.20
	Preservative	q.s.
	Water	65.80
C	Baktol	0.50
D	Sulfur	2.00
	Titanium Dioxide	5.00
	Color, Fragrance	0.50
		100.00

Mixing Procedure:

Combine Part A ingredients and heat to 75-80°C. Mix Part B ingredients and heat to 75-80°C, Add Part C to Part B. Emulsify Parts B+C into Part A. Part D ingredients are finely ground. Slowly stir Part D into A+B+C, and add fragrance.

ANTIACNE CREAM

Supplier: Hüls America, Inc.

	Ingredient	Wt.%
	PEG-3 Stearate (**Marlosol 183**)	5.00
A	Glyceryl Cocoate(and)Hydrogenated Coconut Oil(and)Ceteareth-25 (**Softisan 601**)	4.00
	Caprylic/Capric Triglyceride (**Miglyol 812**)	3.00
	Glyceryl Stearate (**Imwitor 900**)	3.00
	Azelaic Acid	20.00
	Glycerin	1.50
B	Propylene Glycol	12.50
	Preservative	q.s.
	Water	48.00
C	Triethanolamine	3.00
		100.00

Mixing Procedure:

Heat Part A ingredients to 75-80°C. Heat Part B ingredients to 75-80°C. Add Part C to Part B, then emulsify Parts B+C into Part A.

ACNE CREAM

Supplier: Inolex Chemical Company

	Ingredient	Wt.%
	Glyceryl Stearate (**Lexemul 515**)	6.00
A	Isopropyl Myristate (**Lexol IPM**)	3.00
	Stearic Acid	2.00
	Sulfur	2.00
B	Bentonite	5.00
	Alcohol SD40	10.00
	Propylene Glycol	3.00
	Resorcinol	3.00
C	Triethanolamine	1.00
	Methylparaben (**Lexgard M**)	0.15
	Water	64.85
		100.00

Mixing Procedure:

Heat Part C ingredients together to 65°C. Disperse Sulfur and Bentonite in Part C. Melt Part A ingredients together. Add Part A to Part C with agitation. Slowly add Alcohol maintaining good agitation. Cool to 30°C and fill.

BENZOYL PEROXIDE ACNE CREAM

Supplier: R.T Vanderbilt Company, Inc.

	Ingredient	Wt.%
	Magnesium Aluminum Silicate (**Veegum HV**)	2.00
A	Water	38.35
	Hydroxypropyl Methylcellulose	1.50
	Alcohol SDA40	30.00
B	Propylene Glycol	6.00
	Laureth-4	5.00
C	Benzoyl Peroxide, 70%	7.15
	Water	10.00
		100.00

Mixing Procedure:

Add **Veegum HV** to water slowly, agitating continually until smooth. Add the Hydroxypropyl Methylcellulose and mix briefly to obtain a uniform slurry. Add Part B ingredients to Part A and mix until smooth (HPMC dissolved). Mix Part C ingredients until uniform and add to Parts A + B. Pass the cream through a 3-roll mill until smooth and uniform.

5% BENZOYL PEROXIDE LOTION

Supplier: R.T. Vanderbilt Company, Inc.

	Ingredient	Wt.%
	Magnesium Aluminum Silicate (**Veegum**)	0.90
A	Xanthan Gum (**Rhodigel**)	0.40
	Water	75.55
B	Propylene Glycol	6.00
	Benzoyl Peroxide, 70%	7.15
C	Laureth-4	5.00
	Acetylated Lanolin Alcohol	5.00
		100.00

Mixing Procedure:

Add a dry blend of **Veegum** and **Rhodigel** to the water slowly, agitating at maximum available shear until smooth. Mill Part B ingredients until smooth. Blend Part B with Part C and add to Part A. Mix until smooth and uniform.

ACNE TREATMENT LOTION

Supplier: R.T. Vanderbilt Company, Inc.

	Ingredient	Wt.%
	Magnesium Aluminum Silicate (**Veegum**)	0.90
A	Xanthan Gum (**Rhodigel**)	0.40
	Deionized Water	80.70
B	Propylene Glycol	6.00
	Capryloyl Collagen Amino Acids	2.00
C	Laureth-4	5.00
	Acetylated Lanolin Alcohol	5.00
	Preservative	q.s.
		100.00

Mixing Procedure:

Blend **Veegum** and **Rhodigel** and add to water, mixing with maximum available shear until smooth and uniform. Add Part B to Part A. Blend Part C ingredients and heat to 50°C with mixing. Add Part C to Parts A + B with high speed mixing until a uniform emulsion is formed. Cool to 30°C and add preservative.

---

[1] Billow, J. A., *Handbook Of Nonprescription Drugs*, 9th Edition, 799, American Pharmaceutical Association, Washington, DC.

CHAPTER 3

ANTACID/ANTIGAS PRODUCTS

The primary active ingredients used in OTC antacid formulations are Sodium Bicarbonate, Calcium Carbonate, Aluminum Hydroxide, Magnesium Oxide or Hydroxide, Magaldrate (Aluminum Magnesium Hydroxide Sulfate) and Dihydroxy Aluminum Sodium Carbonate. The primary antiflatulant ingredient is Simethicone.

Product forms include liquids, suspensions, tablets, chewable tablets, capsules and caplets.

PART I

BRAND NAME FORMULATIONS

ALKA-MINTS

Manufacturer: Miles, Inc.

Active Ingredient & Concentration:
850mg Calcium Carbonate/Tablet

Product Form:
Tablet, Ch.

Other Ingredients:
Dioctyl Sodium Sulfosuccinate, Flavor, Hydrolyzed Cereal Solids, Magnesium Stearate, Polyethylene Glycol, Sorbitol, Sugar.

ALKA-SELTZER, EFFERVESCENT ANTACID AND PAIN RELIEVER, Regular & Flavored

Manufacturer: Miles, Inc.

Active Ingredients & Concentration:
Regular: 325mg Aspirin
+ 1000mg Citric Acid
+ 1916mg Sodium Bicarbonate/Tablet
Flavored: 325mg Aspirin + 1220mg Citric Acid
+ 1710mg Sodium Bicarbonate/Tablet

Product Form:
Tablet

Other Ingredients:
Regular product has none.
Flavored product has Saccharin Sodium, Flavor

ALKA-SELTZER, EFFERVESCENT ANTACID

Manufacturer: Miles, Inc.

Active Ingredients & Concentration:
958 mg Sodium Bicarbonate
+ 832mg Citric Acid
+ 312mg Potassium Bicarbonate/Tablet

Product Form:
Tablet

Other Ingredient:
Tableting Aid.

ALKA-SELTZER, EXTRA STRENGTH ANTACID & PAIN RELIEVER

Manufacturer: Miles, Inc.

Active Ingredients & Concentration:
500mg Aspirin + 1985mg Sodium Bicarbonate + 1000mg Citric Acid/Tablet

Product Form:
Tablet

Other Ingredients:
Flavors.

ALternaGEL

Manufacturer: Johnson & Johnson • MERCK Consumer Pharmaceuticals Company

Active Ingredient & Concentration:
600mg Aluminum Hydroxide/5ml

Product Form:
Suspension

Other Ingredients:
Butylparaben, Flavor, Propylparaben, Purified Water, Simethicone and other ingredients.

ALUDROX

Manufacturer: Wyeth-Ayerst Laboratories

Active Ingredients & Concentration:
307mg Aluminum Hydroxide Gel
+ 103mg Magnesium Hydroxide/5ml

Product Form:
Suspension

Other Ingredients:
Benzoic Acid, Butylparaben, Glycerin, Hydroxypropyl Methylcellulose, Methylparaben, Propylparaben, Saccharin, Simethicone, Sorbitol, Water, Flavors.

AMPHOJEL

Manufacturer: Wyeth-Ayerst Laboratories

Active Ingredient & Concentration:
Suspension: 320mg Aluminum Hydroxide Gel/5ml
Single Strength Tablet: 300mg Aluminum Hydroxide Gel/Tablet
Double Strength Tablet: 600mg Aluminum Hydroxide Gel/Tablet

Product Form:
Suspension
Tablet

Other Ingredients:
Suspension contains Flavors, Potassium Butylparaben, Calcium Benzoate, Glycerin, Hydroxypropyl Methylcellulose, Potassium Propylparaben, Saccharin, Simethicone, Sorbitol, Peppermint Oil, Menthol, Water. May also contain Flavor, Potassium Butylparaben or Butylparaben
Tablets contain Flavor, Cellulose, Hydrogenated Vegetable Oil, Magnesium Stearate, Polacrilin Potassium, Saccharin, Starch, Talc.

BASAJEL

Manufacturer: Wyeth-Ayerst Laboratories

Active Ingredients & Concentration:
Suspension: Aluminum Carbonate Gel Equal to 400mg Aluminum Hydroxide/5ml
Tablets and Capsules: Aluminum Carbonate Gel Equal to 500mg Aluminum Hydroxide/Tablet or Capsule

Product Forms:
Suspension
Tablet
Capsule

Other Ingredients:
Suspension contains Flavors, Butylparaben, Calcium Benzoate, Glycerin, Hydroxypropyl Methylcellulose, Methylparaben, Mineral Oil, Propylparaben, Saccharin, Simethicone, Sorbitol, Water.
Tablets contain Cellulose, Hydrogenated Vegetable Oil, Magnesium Stearate, Polacrilin Potassium, Starch, Talc.
Capsules contain Colors, Gelatin, Polacrilin Potassium, Polyethylene Glycol, Talc, Titanium Dioxide.

CHOOZ, ANTACID GUM

Manufacturer: Schering-Plough HealthCare Products

Active Ingredient & Concentration: 500mg Calcium Carbonate/Tablet

Product Form: Tablet, Ch.

Other Ingredients:
Sucrose, Gum Base, Glucose, Corn Starch, Peppermint Oil, Hydrated Silica, Gelatin, Glycerin, Acacia, Carnauba Wax, Beeswax, Sodium Benzoate.

CITROCARBONATE

Manufacturer: Roberts Pharmaceutical Corp.

Active Ingredients & Concentration:
57.08% Sodium Bicarbonate
29.02% Citric Acid

Product Form: Powder

Other Ingredients:
6.21% Sodium Citrate, 3.68% Calcium Lactate Pentahydrate, 1.93% Sodium Chloride, 1.07% Monobasic Sodium Phosphate, 1.02% Magnesium Sulfate.

DI-GEL

Manufacturer: Schering-Plough HealthCare Products

Active Ingredients & Concentration:
Tablet: 280mg Calcium Carbonate
+ 128mg Magnesium Hydroxide
+ 20mg Simethicone/Tablet
Suspension: 200mg Aluminum Hydroxide Dried Gel
+ 200mg Magnesium Hydroxide
+ 20 mg Simethicone/5ml

Product Forms:
Tablet
Suspension

Other Ingredients:
Tablets contain D&C Yellow No. 10 Aluminum Lake, Dextrin, FD&C Yellow No. 6 Aluminum Lake, Flavor, Magnesium Stearate, Povidone, Stearic Acid, Mannitol, Sucrose, Talc.
Suspension contains Flavor, Hydroxypropyl Methyl Cellulose, Methylcellulose, Methylparaben, Sorbitol, Propylparaben, Saccharin Sodium, Water.

GAS-X

Manufacturer: Sandoz Pharmaceuticals/Consumer Division

Active Ingredient & Concentration: 80mg Simethicone/Tablet

Product Form: Tablet

Other Ingredients:
Calcium Phosphate Dibasic and Tribasic, Calcium Silicate, Colloidal Silicon Dioxide, Sugar, Microcrystalline Cellulose, Talc, (may also contain Red No. 30, Flavor).

GAS-X, EXTRA STRENGTH

Manufacturer: Sandoz Pharmaceuticals/Consumer Division

Active Ingredient & Concentration: 125mg Simethicone/Tablet

Product Form: Tablet

Other Ingredients:
Calcium Phosphate Dibasic and Tribasic, Calcium Silicate, Colloidal Silicon Dioxide, Sugar, Microcrystalline Cellulose, Talc. May also contain Flavor, Red No. 30, Yellow No.10.

GAVISCON

Manufacturer: SmithKline Beecham Consumer Brands

Active Ingredients & Concentration:
Suspension: 95mg Aluminum Hydroxide + 358mg Magnesium Carbonate/15ml
Tablets: 80mg Aluminum Hydroxide Dried Gel + 20mg Magnesium Trisilicate/Tablet

Product Forms:
Tablet
Suspension

Other Ingredients:
Suspension contains D&C Yellow No. 10, Edetate Disodium, FD&C Blue No. 1, Flavor, Glycerin, Paraben Preservatives, Saccharin Sodium, Sodium Alginate, Sorbitol, Water, Xanthan Gum.
Tablets contain Alginic Acid, Calcium Stearate, Flavor, Sodium Bicarbonate, Starch (may also contain Corn Starch), Sucrose.

GAVISCON, EXTRA STRENGTH

Manufacturer: SmithKline Beecham Consumer Brands

Active Ingredients & Concentration:
Suspension: 508mg Aluminum Hydroxide + 475mg Magnesium Carbonate/10ml
Tablet: 160mg Aluminum Hydroxide + 105mg Magnesium Carbonate/Tablet

Product Forms:
Suspension
Tablet, Ch.

Other Ingredients:
Suspension contains Butylparaben, Edetate Disodium, Flavor, Glycerin, Propylparaben, Saccharin Sodium, Simethicone Emulsion, Sodium Alginate, Sorbitol, Water, Xanthan Gum.
Tablets contain Alginic Acid, Calcium Stearate, Flavor, Mannitol, Sodium Bicarbonate, Stearic Acid, Sucrose.

GAVISCON-2

Manufacturer: SmithKline Beecham Consumer Brands

Active Ingredients & Concentration:
160mg Aluminum Hydroxide Dried Gel + 40mg Magnesium Trisilicate/Tablet

Product Form:
Tablet, Ch.

Other Ingredients:
Alginic Acid, Calcium Stearate, Flavor, Sodium Bicarbonate, Starch (may also contain Corn Starch), Sucrose.

GELUSIL

Manufacturer: Parke-Davis

Active Ingredients & Concentration:
200mg Aluminum Hydroxide + 200 mg Magnesium Hydroxide + 25mg Simethicone/5ml Of Suspension or Tablet

Product Forms:
Suspension
Tablet

Other Ingredients:
Suspension contains Ammonium Hydroxide, Calcium Hypochlorite, Citric Acid, Flavors, Hydroxypropyl Methylcellulose, Menthol, Saccharin Sodium, Sorbitol, Water, Xanthan Gum.
Tablets contain Flavors, Magnesium Stearate, Mannitol, Sorbitol, Sugar.

MAALOX

Manufacturer: Rhône-Poulenc Rorer Pharmaceuticals,Inc.

Active Ingredients & Concentration:
Caplets: 1000mg Calcium Carbonate/Caplet
Suspension: 200mg Magnesium Hydroxide + 225mg Aluminum Hydroxide/5ml

Product Forms:
Caplet
Suspension

Other Ingredients:
Caplets contain Corn Starch, Croscarmellose, Magnesium Stearate, Sodium Lauryl Sulfate.
Suspension contains Citric Acid, Flavors, Methylparaben, Propylparaben, Saccharin Sodium, Sorbitol, Purified Water and other ingredients.

MAALOX, PLUS

Manufacturer: Rhône-Poulenc Rorer Pharmaceuticals,Inc.

Active Ingredients & Concentration:
200mg Aluminum Hydroxide
+ 200mg Magnesium Hydroxide
+ 25mg Simethicone/Tablet

Product Form:
Tablet

Other Ingredients:
Citric Acid, Sugar, Colors, Dextrose, Flavors, Glycerin, Magnesium Stearate, Mannitol, Saccharin Sodium, Sorbitol, Starch, Talc (may also contain Hydrogenated Vegetable Oil).

MAALOX, GRF

Manufacturer: Rhône-Poulenc Rorer Pharmaceuticals,Inc.

Active Ingredient & Concentration:
80mg Simethicone/Tablet

Product Form:
Tablet

Other Ingredients:
Calcium Stearate, Corn Starch, D&C Red No. 27 Aluminum Lake, Flavor, Gelatin, Mannitol, Sucrose, Tribasic Calcium Phosphate.

MAALOX, EXTRA STRENGTH PLUS

Manufacturer: Rhône-Poulenc Rorer Pharmaceuticals,Inc.

Active Ingredients & Concentration:
Suspension: 500mg Aluminum Hydroxide + 450mg Magnesium Hydroxide + 40mg Simethicone/5ml
Tablet: 350mg Aluminum Hydroxide + 350mg Magnesium Hydroxide + 30mg Simethicone/Tablet

Product Forms:
Suspension
Tablet

Other Ingredients:
Suspension contains Citric Acid, Flavors, Methylparaben, Propylparaben, Purified Water, Saccharin Sodium, Sorbitol and other ingredients.
Tablets contain Citric Acid, Sugar, D&C Yellow No. 10, Dextrose, FD&C Blue No. 1, Flavors, Glycerin, Hydrogenated Vegetable Oil, Magnesium Stearate, Mannitol, Saccharin Sodium, Sorbitol, Starch, Talc.

MAALOX, HRF

Manufacturer: Rhône-Poulenc Rorer Pharmaceuticals,Inc.

Active Ingredients & Concentration:
Suspension: 280mg Aluminum Hydroxide-Magnesium Carbonate Codried Gel + 350mg Magnesium Carbonate/10ml
Tablet: 180mg Aluminum Hydroxide-Magnesium Carbonate Codried Gel + 160mg Magnesium Carbonate/Tablet

Product Forms:
Suspension
Tablet

Other Ingredients:
Suspension contains Calcium Carbonate, Saccharin Calcium, FD&C Blue No. 1, FD&C Yellow No. 5, Flavors, Magnesium Alginate, Methylparaben, Propylparaben, Potassium Bicarbonate, Purified Water, Sorbitol and other ingredients.
Tablets contain Sugar, Corn Starch, D&C Yellow No. 10, FD&C Blue No. 2, Flavors, Magnesium Alginate, Magnesium Stearate, Potassium Bicarbonate.

MYLANTA

Manufacturer: Johnson & Johnson • MERCK Consumer Pharmaceuticals Co.

Active Ingredients & Concentration:
311mg Calcium Carbonate
+ 232mg Magnesium Carbonate/Gelcap

Product Form:
Gelcap

Other Ingredients:
Benzyl Alcohol, Butylparaben, Castor Oil, D&C Yellow No. 10, Edetate Disodium Calcium, FD&C Blue No. 1, Gelatin, Hydroxypropylcellulose, Magnesium Stearate, Methylparaben, Microcrystalline Cellulose, Propylparaben, Croscarmellose Sodium, Sodium Lauryl Sulfate, Sodium Propionate, Titanium Dioxide.

MYLANTA, ANTACID/ANTIGAS

Manufacturer: Johnson & Johnson • MERCK Consumer Pharmaceuticals Co.

Active Ingredients & Concentration:
200mg Aluminum Hydroxide
+ 200mg Magnesium Hydroxide
+ 20mg Simethicone/5ml of Suspension or Chewable Tablet

Product Forms:
Suspension
Tablet, Ch.

Other Ingredients:
Suspension contains Butylparaben, Flavors, Carboxymethylcellulose Sodium, Propylparaben, Hydroxypropyl Methylcellulose, Microcrystalline Cellulose, Purified Water, Saccharin Sodium, Sorbitol.
Tablets contain Dextrates, Flavors, Magnesium Stearate, Mannitol, Sorbitol Colloidal Silicon Dioxide, Saccharin Sodium, Colors.

MYLANTA, DOUBLE STRENGTH

Manufacturer: Johnson & Johnson • MERCK Consumer Pharmaceuticals Co.

Active Ingredients & Concentration:	Product Forms:
400mg Aluminum Hydroxide + 400mg Magnesium Hydroxide + 40mg Simethicone/5 ml of Suspension or Chewable Tablet	Suspension Tablet, Ch.

Other Ingredients:

Suspension contains Butylparaben, Flavors, Carboxymethylcellulose Sodium, Propylparaben, Hydroxypropyl Methylcellulose, Microcrystalline Cellulose, Potassium Citrate, Purified Water, Saccharin Sodium, Sorbitol.

Tablets contain Colloidal Silicon Dioxide, Dextrates, Flavors, Magnesium Stearate, Mannitol, Saccharin Sodium, Sorbitol, Colors.

MYLANTA, GAS

Manufacturer: Johnson & Johnson • MERCK Consumer Pharmaceuticals Co.

Active Ingredients & Concentration:	Product Form:
40, 80 or 125mg Simethicone/Tablet	Tablet

Other Ingredients:

40mg tablets contain Calcium Silicate, Lactose, Povidone, Saccharin Calcium.

80 and 125mg tablets contain Dextrates., Flavor, Sorbitol, Stearic Acid, Tricalcium Phosphate.

MYLICON

Manufacturer: Johnson & Johnson • MERCK Consumer Pharmaceuticals Co.

Active Ingredient & Concentration:	Product Form:
40mg Simethicone/0.6 ml	Drops

Other Ingredients:

Carbomer 934, Citric Acid, Flavors, Hydroxypropyl Methylcellulose, Purified Water, Red No. 3, Saccharin Calcium, Sodium Benzoate, Sodium Citrate.

NEPHROX

Manufacturer: Fleming & Company

Active Ingredient & Concentration:	Product Form:
320mg Aluminum Hydroxide as Gel/5ml	Liquid

Other Ingredients:
Mineral Oil(10% by volume), Flavor.

PHAZYME

Manufacturer: Reed & Carnrick

Active Ingredients & Concentration:	Product Forms:
Tablet: 60mg Simethicone/Tablet	Tablet
Liquid: 40mg Simethicone/0.6ml	Liquid

Other Ingredients::
Tablets contain Acacia. Calcium Sulfate, Carnauba Wax, Crospovidone, Colors, Gelatin, Lactose, Methylparaben, Microcrystalline Cellulose, PEG-40 Stearate, Povidone, Pregelatinized Starch, Propylparaben, Rice Starch, Sodium Benzoate, Sucrose, Talc, Titanium Dioxide, White Wax.
Liquid contains Carbomer 934P, Citric Acid, Flavor, Hydroxypropyl Methylcellulose, PEG-8 Stearate, Potassium Sorbate, Sodium Citrate, Saccharin Sodium, Water.

PHAZYME 95

Manufacturer: Reed & Carnrick

Active Ingredient & Concentration:	Product Form:
95mg Simethicone/Tablet	Tablet

Other Ingredients:
Tablets contain Acacia, Calcium Sulfate, Carnauba Wax, Crospovidone, Colors, Gelatin, Lactose, Methylparaben, Microcrystalline Cellulose, PEG-40 Stearate, Povidone, Pregelatinized Starch, Propylparaben, Rice Starch, Sodium Benzoate, Sucrose, Talc, Titanium Dioxide, White Wax.

PHAZYME 125

Manufacturer: Reed & Carnrick

Active Ingredient & Concentration: 125mg Simethicone/Tablet

Product Form: Gelcap

Other Ingredients:
Gelatin, Glycerin, Hydrogenated Soybean Oil, Lecithin, Methylparaben, Polysorbate 80, Propylparaben, Soybean Oil, Titanium Dioxide, Vegetable Shortening, Yellow Wax.

RIOPAN

Manufacturer: Whitehall Laboratories, Inc.

Active Ingredients & Concentration: 540mg Magaldrate/5 ml

Product Form: Suspension

Other Ingredients:
Flavor, Glycerin, Potassium Citrate, Saccharin, Sorbitol, Xanthan Gum, Water.

RIOPAN, PLUS

Manufacturer: Whitehall Laboratories, Inc.

Active Ingredients & Concentration: 540mg Magaldrate + 40mg Simethicone/5 ml

Product Form: Suspension

Other Ingredients:
Flavor, Glycerin, PEG-8 Stearate, Potassium Citrate, Saccharin, Sorbitan Stearate, Sorbitol, Xanthan Gum, Water.

RIOPAN, PLUS 2

Manufacturer: Whitehall Laboratories, Inc.

Active Ingredient & Concentration:
1080mg Magaldrate
+ 40mg Simethicone/5ml

Product Form:
Suspension

Other Ingredients:
Flavor, Glycerin, PEG-8 Stearate, Potassium Citrate, Saccharin, Sorbitan Stearate, Sorbitol, Xanthan Gum, Water.

ROLAIDS

Manufacturer: Warner-Lambert Company

Active Ingredient & Concentration:
300mg Dihydroxyaluminum Sodium Carbonate/Tablet

Product Form:
Tablet

Other Ingredients:
Corn Starch, Corn Syrup, Flavor, Magnesium Stearate, Sugar, (may also contain Pregelatinized Starch).

ROLAIDS, CALCIUM RICH/SODIUM FREE

Manufacturer: Warner-Lambert Company

Active Ingredient & Concentration:
550mg Calcium Carbonate/Tablet

Product Form:
Tablet

Other Ingredients:
Corn Starch, Flavor, Magnesium Stearate, Mineral Oil, Mannitol, Color, Silicon Dioxide, Sucrose, Pregelatinized Starch.

ROLAIDS, EXTRA STRENGTH

Manufacturer: Warner-Lambert Company

Active Ingredients & Concentration:
1000mg Calcium Carbonate
+ 25mg Simethicone/Tablet

Product Form:
Tablet

Other Ingredients:
Acesulfame Potassium, FD&C Blue No. 1, FD&C Yellow No. 5, Titanium Dioxide, Corn Syrup, Dextrose, Glyceryl Monostearate, Flavor, Pregelatinized Starch, Sucrose, Triglycerol Monooleate.

TUMS

Manufacturer: SmithKline Beecham Consumer Brands

Active Ingredient & Concentration:
500mg Calcium Carbonate/Tablet

Product Form:
Tablet

Other Ingredients:
Flavor, Mineral Oil, Sodium Polyphosphate, Starch, Sucrose, Talc.(may also contain Adipic Acid, FD&C Yellow No. 6).

TUMS, E-X

Manufacturer: SmithKline Beecham Consumer Brands

Active Ingredient & Concentration:
750mg Calcium Carbonate/Tablet

Product Form:
Tablet

Other Ingredients:
Flavor, Mineral Oil, Sodium Polyphosphate, Starch, Sucrose, Talc. (may also contain Adipic Acid, FD&C Yellow No. 6).

TUMS PLUS, ANTACID/ANTIGAS

Manufacturer: SmithKline Beecham Consumer Brands

Active Ingredients & Concentration:
500mg Calcium Carbonate
+ 20mg Simethicone/Tablet

Product Form:
Tablet

Other Ingredients:
Adipic Acid Corn Syrup, Microcrystalline Cellulose, Flavors, Mineral Oil, Sodium Polyphosphate, Starch, Sucrose, Talc, Triglyceryl Monooleate, Colors.

WINGEL

Manufacturer: Sterling Winthrop, Inc.

Active Ingredients & Concentration:
180mg Aluminum Hydroxide
+ 160mg Magnesium Hydroxide/5 ml

Product Form:
Suspension

Other Ingredients:
Benzoic Acid, Flavor, Methylcellulose, Purified Water, Iron Oxide, Saccharin Sodium, Sodium Hypochlorite, Sorbitol.

PART II

SUPPLIER'S FORMULATIONS

ANTACID

Supplier: American Colloid Company

	Ingredient:	Wt.%
A	Magnesium Aluminum Silicate (**Magnabrite S**) 5% Dispersion	20.00
B	Methylparaben	0.20
C	Propylparaben	0.10
D	Saccharin Sodium	0.05
E	Aluminum Hydroxide-Magnesium Hydroxide Fluid Gel	50.00
F	Polysorbate 80	0.30
G	Flavor	0.20
H	Deionized Water	29.15
		100.00

Mixing Procedure:

Add the Parabens and Saccharin to the **Magnabrite S** dispersion with stirring until they dissolve. Add Part E with mixing. Disperse Part G in Part F and add to the batch.

CALCIUM EFFERVESCENT TABLETS

Supplier: BASF Corp.

	Ingredient:	Wt.%
A	Calcium Lactate	25.36
	Calcium Gluconate	24.39
	Calcium Carbonate	7.41
	Sodium Bicarbonate	16.00
	Tartaric Acid	18.73
	Povidone (**Kollidon 30**)	1.87
B	Povidone (**Kollidon 30**)	0.47
	Isopropyl or Ethyl Alcohol, 96%	q.s.
C	Crospovidone (**Kollidon CL**)	3.90
	PEG 6000	1.87
	Flavor	q.s.
		100.00

Mixing Procedure:

Granulate the Part A ingredients with Part B solution, dry, sieve, mix well with Part C ingredients and press with low compression force to form tablets of 2,500mg weight and 20mm diameter.

MAGALDRATE ANTACID SUSPENSION

Supplier: BASF Corp.

Ingredient:	
Magaldrate	100.0gm
Crospovidone (**Kollidon CL-M**)	80.0gm
Povidone (**Kollidon 90F**)	30.0gm
Sucrose	150.0gm
Flavors	10.3gm
Saccharin Sodium	0.2gm
Preservative	q.s.
Water	to 1000.0ml

Mixing Procedure:

Dissolve or suspend the solids in the water under aseptic conditions.

MAGALDRATE CHEWABLE TABLET

Supplier: BASF Corp.

	Ingredient:	Wt.%
	Magaldrate	42.70
A	Lactose Monohydrate	34.16
	Flavor I	4.27
	Povidone (**Kollidon 90F**)	1.71
B	Flavors II	1.02
	Saccharin Sodium	0.09
	Water	15.36
C	Silicon Dioxide	0.43
	Magnesium Stearate	0.26
		100.00

Mixing Procedure:

Granulate the Part A ingredients with the Part B solution, pass through a 0.8mm sieve, dry, mix with the Part C ingredients and press with low compressive force to tablets weighing 1,000mg, diameter of 16mm.

ANTACID SUSPENSION

Supplier: FMC Corp.

Ingredient:	Wt.%
Aluminum Hydroxide Gel	24.46
Magnesium Hydroxide Paste	12.94
Sorbitol, 70%	5.00
Methylparaben	0.10
Propylparaben	0.01
Microcrystalline Cellulose (**Avicel® RC-591, NF**)	0.90
Xanthan Gum	0.10
Water	56.49
	100.00

Mixing Procedure:

Measure total volume of water and add preservatives. Heat, if necessary, to dissolve more quickly. Mix until completely dissolved. Slowly add **Avicel RC-591** while mixing. Agitate until Avicel is fully dispersed. Add Xanthan Gum to the dispersion slowly, while mixing. Mix until dissolved. Add Sorbitol solution and mix. Slowly add Aluminum Hydroxide gel and mix until homogeneous. Slowly add Magnesium Hydroxide paste and mix until homogeneous.

SIMETHICONE TABLET, 50 MG/Tablet

Supplier: FMC Corp.

Ingredient:	Wt.%
Simethicone	7.14
Microcrystalline Cellulose (**Avicel PH-105**)	7.14
Magnesium Hydroxide	7.14
Mannitol	26.50
Lactose	10.00
Granular Sugar	39.51
Menthol	0.07
Fumed Silica (Syloid-266)	1.00
Fumed Silica (Syloid-74)	0.50
Magnesium stearate	1.00
	100.00

Mixing Procedure:

Blend **Avicel**[R] **PH-105** and Magnesium Hydroxide in a V-shaped blender for 10 minutes. Remove this blend from blender and place in a planetary mixer. Slowly add the weighed amount of Simethicone to the mix using a 'B' flat beater blade with mixer on slow speed. After Simethicone is thoroughly mixed pass through 20 mesh screen. A formulation can now be made using the above mixture and the remaining ingredients in the specified proportions.

ALUMINUM HYDROXIDE/MAGNESIUM HYDROXIDE SUSPENSION

SINGLE STRENGTH: 4.6% $Al(OH)_3$, 2.9% $Mg(OH)_2$

Supplier: Giulini Corp.

Ingredient:	Wt.%
Aluminum Hydroxide Gel (**Alugel A661**)	31.60
Magnesium Hydroxide Paste (**Gilumag D661**)	9.60
Glycerin	3.00
Methylcellulose, 6000 cPs. Viscosity	0.20
Sorbitol	5.00
Methylparaben	0.10
Propylparaben	0.05
Flavor, Sweeteners	1.00
Water	49.45
	100.00

Mixing Procedure:

Contact supplier for procedure.

ALUMINUM HYDROXIDE/MAGNESIUM HYDROXIDE SUSPENSION

DOUBLE STRENGTH: 9.2% $Al(OH)_3$, 4.8% $Mg(OH)_2$

Supplier: Giulini Corp.

Ingredient:	Wt.%
Aluminum Hydroxide Wet Gel (**Alugel A661**)	63.00
Magnesium Hydroxide Gel (**Gilumag D611**)	14.50
Sorbitol	15.00
Glycerin	2.00
Methylparaben	0.10
Propylparaben	0.05
Flavor, Sweetener	1.00
Water	4.35
	100.00

Mixing Procedure:
Contact supplier for procedure.

ALUMINUM HYDROXIDE/MAGNESIUM HYDROXIDE SUSPENSION

TRIPLE STRENGTH: 10.7% $Al(OH)_3$, 7.2% $Mg(OH)_2$

Supplier: Giulini Corp.

Ingredient:	Wt.%
Aluminum Hydroxide Wet Gel (**Alugel A671**)	56.00
Magnesium Hydroxide Gel (**Gilumag D611**)	24.00
Sorbitol	10.00
Methylcellulose, 4000 cPs. viscosity	0.10
Glycerin	1.00
Simethicone	1.00
Methylparaben	0.10
Propylparaben	0.05
Flavor, Sweetener	1.00
Water	6.75
	100.00

Mixing Procedure:
Contact supplier for procedure.

ALUMINUM HYDROXIDE/MAGNESIUM HYDROXIDE SUSPENSION

SINGLE STRENGTH: 3.4% $Al(OH)_3$, 5.2% $Mg(OH)_2$

Supplier: Giulini Corp.

Ingredient:	Wt.%
Aluminum Hydroxide Wet Gel (**Alugel A681**)	13.00
Magnesium Hydroxide Paste (**Gilumag D611**)	12.00
Methylcellulose	1.00
Glycerin	2.00
Sorbitol	20.00
Methylparaben	0.10
Propylparaben	0.05
Flavor, Sweetener	1.00
Water	50.85
	100.00

Mixing Procedure:
Contact supplier for procedure.

ALUMINUM HYDROXIDE/MAGNESIUM HYDROXIDE SUSPENSION

DOUBLE STRENGTH: 6.8% $Al(OH)_3$, 10.4% $Mg(OH)_2$

Supplier: Giulini Corp.

Ingredient:	Wt.%
Aluminum Hydroxide Wet Gel (**Alugel A681**)	26.00
Magnesium Hydroxide Gel(**Gilumag D611**)	24.00
Methylcellulose	1.00
Glycerin	2.00
Sorbitol	20.00
Methylparaben	0.10
Propylparaben	0.05
Flavor, Sweetener	1.00
Water	25.85
	100.00

Mixing Procedure:
Contact supplier for procedure.

MAGALDRATE SUSPENSION (8%)

Supplier: Giulini Corp.

Ingredient:	Wt.%
Magaldrate (**Giulini C631**)	37.50
Sorbitol	7.50
Glycerin	1.00
Carboxymethylcellulose, 4000 cPs. Visc.	2.50
Methylparaben	0.10
Propylparaben	0.05
Flavor, Sweetener	1.00
Water	50.35
	100.00

Mixing Procedure:
Contact supplier for procedure.

HYDROTALCITE TABLET

Supplier: Giulini Corp.

Ingredient:	Wt.%
Hydrotalcite (**Giulini C300**)	61.00
Mannitol	31.00
Corn Starch	3.00
Talc	2.00
Magnesium Stearate	2.00
Flavor, Sweetener	1.00
	100.00

Mixing Procedure:
Contact supplier for procedure.

MAGALDRATE TABLET

Supplier: Giulini Corp.

Ingredient:	Wt.%
Magaldrate Powder (**Giulini C410**)	67.00
Mannitol	21.00
Microcrystalline Cellulose	5.00
Magnesium Stearate	2.00
Simethicone	4.00
Flavor, Sweetener	1.00
	100.00

Mixing Procedure:
Contact supplier for procedure.

DASC TABLET

Supplier: Giulini Corp.

Ingredient:	Wt.%
Dihydroxyaluminum Sodium Carbonate (**Giulini A265**)	31.00
Sugar	61.50
Magnesium Stearate	2.00
Starch	5.00
Flavor, Sweetener	0.50
	100.00

Mixing Procedure:

Contact supplier for procedure.

CHEWABLE ANTACID TABLETS

Supplier: International Specialty Products

	Ingredient:	Wt.%
	Aluminum Hydroxide (Dried Gel)	32.04
A	Magnesium Hydroxide Powder	32.04
	Sucrose	3.20
	Mannitol	28.84
	Povidone, (**Plasdone®**), 10% Solution	
B	in 50% Water/50% Alcohol	q.s.
	Glycerin	1.29
	Magnesium Stearate	1.92
C	Fumed Silica	0.64
	Oil of Peppermint	0.03
		100.00

Mixing Procedure:

Mix the Part A ingredients in a suitable blender. Add Glycerin to the **Plasdone** solution and use the combination to moisten the Part A ingredients.

Granulate Parts A + B by passing through a 14 mesh screen. Dry at 140-150°F. Mix Part C ingredients and size through a 20-mesh screen. Mix Parts A, B and C thoroughly and compress using 0.5 inch, flat face, beveled-edge punches.

ALUMINA + MAGNESIA ANTACID I
3.4% ALUMINUM HYDROXIDE, 1.7% MAGNESIUM HYDROXIDE

Supplier: Kelco Division of Merck & Company, Inc.

Ingredient:	Wt.%
Deionized Water	38.85
Aluminum Hydroxide Gel (Liquigel)	40.00
Sorbitol, 70%	15.00
Magnesium Hydroxide (Hydro Magma)	5.50
Xanthan Gum (**Keltrol**)	0.20
Saccharin Sodium	0.20
Methylparaben	0.15
Potassium Citrate	0.10
Flavor	q.s.
	100.00

Mixing Procedure:

Mix Water and Sorbitol for 2 minutes with a propeller mixer. Add the **Keltrol**. Mix 5 minutes. Add Potassium Citrate, Saccharin Sodium and Methylparaben. Mix 5 minutes. Add Aluminum Hydroxide. Mix 2 minutes. Add Magnesium Hydroxide. Mix 2 minutes. Add the Flavor and mix 10 minutes.

ALUMINA + MAGNESIA ANTACID II
2.1% ALUMINUM HYDROXIDE, 3.4% MAGNESIUM HYDROXIDE

Supplier: Kelco Division of Merck & Company, Inc.

Ingredient:	Wt.%
Deionized Water	48.40
Aluminum Hydroxide Gel (Liquigel)	25.00
Sorbitol, 70%	15.00
Magnesium Hydroxide (Hydro Magma)	11.00
Saccharin Sodium	0.20
Xanthan Gum (**Keltrol**)	0.15
Methylparaben	0.15
Potassium Citrate	0.10
Flavor	q.s.
	100.00

Mixing Procedure:

Mix Water and Sorbitol for 2 minutes with a propeller mixer. Add the **Keltrol**. Mix 5 minutes. Add Potassium Citrate, Saccharin Sodium and Methylparaben. Mix 5 minutes. Add Aluminum Hydroxide. Mix 2 minutes. Add Magnesium Hydroxide. Mix 2 minutes. Add Flavor and mix 10 minutes.

ALUMINUM-MAGNESIUM HYDROXIDE SUSPENSION WITH ANTIGAS, DOUBLE STRENGTH

Supplier: Reheis, Inc.

	Ingredient:	Wt.%
A	Aluminum Hydroxide Gel (**Liquigel HO**)	28.08
B	**Magnesium Hydroxide Paste**	24.58
C	Deionized Water	39.17
D	Simethicone Emulsion, 30%	2.42
E	Sorbitol, 70% Solution	4.90
F	Hydroxypropyl Methylcellulose, Medium Visc.	0.65
G	Methylparaben	0.16
H	Propylparaben	0.04
I	Flavor	q.s.
		100.00

Mixing Procedure:

Combine Part D with 50% of Part C using a colloid mill. Mix 5 min. Add Part A and mix 5 min. Add Part B and mix 15 min. Heat 50% of Part C to 80°C and add Parts G and H. Mix thoroughly. Add to batch and mix 5 min. Transfer to overhead mixer. Combine Parts E and F, mix thoroughly and add to the batch while mixing. Add Part I and mix 5 min.

ALUMINUM-MAGNESIUM HYDROXIDE SUSPENSION WITH ANTIGAS, SINGLE STRENGTH

Supplier: Reheis, Inc.

	Ingredient:	Wt.%
A	Aluminum Hydroxide Gel (**Liquigel HO**)	14.04
B	**Magnesium Hydroxide Paste**	12.29
C	Deionized Water	66.62
D	Simethicone Emulsion, 30%	1.30
E	Sorbitol, 70% Solution	4.90
F	Hydroxypropyl Methylcellulose, Medium Visc.	0.65
G	Methylparaben	0.16
H	Propylparaben	0.04
I	Flavor	q.s.
		100.00

Mixing Procedure:

Combine Part D with 75% of Part C using a colloid mill. Mix 5 min. Add Part A and mix 5 min. Add Part B and mix 15 min. Heat 25% of Part C to 80°C and add Parts G and H. Mix thoroughly. Add to batch and mix 5 min. Transfer to overhead mixer. Combine Parts E and F, mix thoroughly and add to the batch while mixing. Add Part I and mix 5 min.

MAGALDRATE ANTACID, EXTRA STRENGTH FORMULA

Supplier: Reheis, Inc.

	Ingredient:	Wt.%
A	**Magaldrate Fluid Gel**	48.40
B	Sorbitol, 70% Solution	7.00
C	Hydroxypropyl Methylcellulose, Medium Visc.	0.50
D	Methylparaben	0.22
E	Propylparaben	0.04
F	Flavor	0.30
G	Deionized Water	43.54
		100.00

Mixing Procedure:

Add Parts A and G to a colloid mill and mix 15 min. Using an overhead mixer, add Part C to Part B and mix 10 min. or until C is thoroughly hydrated. Add Parts C + B to Parts A + G and mix with an overhead mixer. Combine Parts D, E and F. Stir until dissolved and add to the batch. Mix 5 min. with an overhead mixer.

ALUMINUM-MAGNESIUM HYDROXIDE SUSPENSION WITH ANTIGAS, REGULAR STRENGTH

Supplier: Reheis, Inc.

	Ingredient:	Wt.%
A	Aluminum Hydroxide Gel (**Liquigel HO**)	15.80
B	**Magnesium Hydroxide Paste**	12.29
C	Deionized Water	64.86
D	Simethicone Emulsion, 30%	1.30
E	Sorbitol, 70% Solution	4.90
F	Hydroxypropyl Methylcellulose, Medium Visc.	0.65
G	Methylparaben	0.16
H	Propylparaben	0.04
I	Flavor	q.s.
		100.00

Mixing Procedure:

Combine Part D with 75% of Part C using a colloid mill. Mix 5 min. Add Part A and mix 5 min. Add Part B and mix 15 min. Heat 25% of Part C to 80°C and add Parts G and H. Mix thoroughly. Add to batch and mix 5 min. Transfer to overhead mixer. Combine Parts E and F, mix thoroughly and add to the batch while mixing. Add Part I and mix 5 min.

ALUMINUM-MAGNESIUM HYDROXIDE SUSPENSION WITH ANTIGAS, EXTRA STRENGTH

Supplier: Reheis, Inc.

	Ingredient:	Wt.%
A	Aluminum Hydroxide Gel (**Liquigel HO**)	33.51
B	**Magnesium Hydroxide Paste**	26.39
C	Deionized Water	31.93
D	Simethicone Emulsion, 30%	2.42
E	Sorbitol, 70% Solution	4.90
F	Hydroxypropyl Methylcellulose, Medium Visc.	0.65
G	Methylparaben	0.16
H	Propylparaben	0.04
I	Flavor	q.s.
		100.00

Mixing Procedure:

Combine Parts D and C using a colloid mill. Mix 5 minutes. Add Part A and mix 5 min. Add Part B and mix 15 minutes. Transfer to an overhead mixer. Combine Parts E and F, mix thoroughly and add to the batch while mixing. Combine Parts G, H and I and mix until dissolved. Add to the batch and mix 5 minutes.

ANTACID SUSPENSION I

Supplier: R.T. Vanderbilt Company, Inc.

	Ingredient:	Wt.%
	Purified Bentonite (**Veegum HS**)	0.50
A	Xanthan Gum (**Rhodigel**)	0.20
	Water	40.10
	Sorbitol, 70%	20.00
B	Aluminum Hydroxide Gel (AHLT-LW)	36.00
	Magnesium Hydroxide, USP	3.20
C	Preservative, Flavor	q.s.
		100.00

Mixing Procedure:

Add a dry blend of **Veegum HS** and **Rhodigel** to the water slowly, agitating with maximum available shear until smooth and uniform. Mix Part B until uniform and add to Part A. Mix until smooth and uniform. Add Part C and mix until uniform.

ANTACID SUSPENSION II

Supplier: R.T. Vanderbilt Company, Inc.

	Ingredient:	Wt.%
	Purified Bentonite (**Veegum HS**)	0.25
A	Xanthan Gum (**Rhodigel**)	0.28
	Deionized Water	56.02
B	Methylparaben	0.10
	Propylparaben	0.05
C	Sorbitol, 70%	14.30
D	Aluminum Hydroxide Wet Gel, (Alugel A671)	17.00
	Magnesium Hydroxide Paste (Gilumag D611)	12.00
		100.00

Mixing Procedure:

Dry blend **Veegum HS** and **Rhodigel** and add to water, mixing with maximum available shear until smooth and uniform and free of undispersed particles. Add Part B, C and D ingredients in the order shown, mixing each until dissolved or dispersed before adding the next.

CHAPTER 4

ANTIDIARRHEAL PRODUCTS

Bismuth Subsalicylate, Loperamide Hydrochloride and activated Attaplugite clay are the primary active ingredients in OTC antidiarrheral products.

They are marketed in liquid, tablet and/or suspension forms.

PART I

BRAND NAME FORMULATIONS

DIASORB

Manufacturer: Colombia Laboratories, Inc.

Active Ingredient & Concentration:
750mg Activated Attapulgite/5ml of Suspension or Tablet

Product Forms:
Suspension
Tablet

Other Ingredients:
Suspension contains Benzoic Acid, Citric Acid, Flavor, Glycerin, Magnesium Aluminum Silicate, Methylparaben, Polysorbate, Propylene Glycol, Propylparaben Saccharin, Sodium Hypochlorite, Sorbitol, Xanthan Gum, Water.
Tablets contain D&C Red No. 30 Aluminum Lake, Gelatin, Hydroxypropyl Methylcellulose, Magnesium Stearate, Pharmaceutical Shellac, Polyethylene Glycol, Povidone, Propylene Glycol, Sorbitol, Titanium Dioxide, Water.

DONNAGAL

Manufacturer: Wyeth-Ayerst Laboratories

Active Ingredient & Concentration:	Product Forms:
600mg Attapulgite USP/15ml of Suspension or Tablet	Suspension Tablet, Ch.

Other Ingredients:

Suspension contains Alcohol (1.4%), Benzyl Alcohol, Carboxymethylcellulose Sodium, Citric Acid, FD&C Blue No. 1, Flavors, Magnesium Aluminum Silicate, Methylparaben, Phosphoric Acid, Propylene Glycol, Propylparaben, Saccharin Sodium, Sorbitol, Titanium Dioxide, Water, Xanthan Gum.

Tablets contain D&C Yellow No. 10 Aluminum Lake, FD&C Blue No. 1 Aluminum Lake, Flavors, Magnesium Stearate, Mannitol, Saccharin Sodium, Sorbitol, Water.

IMODIUM A-D

Manufacturer: McNeil Consumer Products Company

Active Ingredients & Concentration:	Product Forms:
Liquid: 1.0mg Loperamide Hydrochloride/5 ml Caplet: 2.0mg Loperamide Hydrochloride/Caplet	Liquid Caplet

Other Ingredients:

Liquid contains Alcohol (5.25%), Citric Acid, Flavors, Glycerin, Methylparaben, Propylparaben, Purified Water.

Caplets contain Corn Starch, Lactose, Magnesium Stearate, Microcrystalline Cellulose, FD&C Blue No. 1, D&C Yellow No. 10.

KAOPECTATE

Manufacturer: The Upjohn Company

Active Ingredient & Concentration:
600mg Attapulgite/15ml

Product Form:
Suspension

Other Ingredients:
Flavors, Glucono Delta Lactone, Magnesium Aluminum Silicate, Methylparaben, Sorbic Acid, Sucrose, Titanium Dioxide, Xanthan Gum, Purified Water, Flavors. May also contain FD&C Red No. 40.

KAOPECTATE, MAXIMUM STRENGTH

Manufacturer: The Upjohn Company

Active Ingredient & Concentration:
750mg Attapulgite/Caplet

Product Form:
Caplet

Other Ingredients:
Croscarmellose Sodium, Hydroxypropylcellulose, Hydroxypropyl Methylcellulose, Methylparaben, Pectin, Propylene Glycol, Propylparaben, Sucrose, Titanium Dioxide, Zinc Stearate.

KAOPECTATE, CHILDREN'S STRENGTH

Manufacturer: The Upjohn Company

Active Ingredient & Concentration:
300mg Attapulgite/7.5ml of Suspension or Tablet

Product Forms:
Suspension
Tablet, Ch.

Other Ingredients:
Suspension contains Flavors, Glucono Delta Lactone, Magnesium Aluminum Silicate, Methylparaben, Sorbic Acid, Sucrose, Titanium Dioxide, Xanthan Gum, Purified Water, FD&C Red No. 40.
Tablets contain Corn Starch, Dextrins, Dextrose, D&C Red No. 27, D&C Red No. 30, Flavor, Magnesium Stearate, Sucrose, Titanium Dioxide.

LOPERAMIDE HYDROCHLORIDE

Manufacturer: Ohm Laboratories, Inc.

Active Ingredient & Concentration: 2mg Loperamide Hydrochloride/Caplet

Product Form: Caplet

Other Ingredients:
Croscarmellose Sodium, Crospovidone, Hydrogenated Vegetable Oil, Lactose, Magnesium Stearate, Powdered Cellulose, Pregelatinized Starch, FD&C Blue No. 1, D&C Yellow No. 10.

PEPTO-BISMOL

Manufacturer: Procter & Gamble Company

Active Ingredient & Concentration: 262mg Bismuth Subsalicylate/ 15ml of Suspension or Tablet

Product Forms: Suspension, Tablet

Other Ingredients:
Suspension contains Benzoic Acid, Flavor, Colors, Magnesium Aluminum Silicate, Methyl Cellulose, Saccharin Sodium, Salicylic Acid, Sodium Salicylate, Sorbic Acid, Water.
Tablets contain Calcium Carbonate, Colors, Flavors, Magnesium Stearate, Mannitol, Povidone, Saccharin Sodium, Talc. May also contain Adipic Acid,

---

PEPTO-BISMOL, MAXIMUM STRENGTH

Manufacturer: Procter & Gamble Company

Active Ingredient & Concentration: 525mg Bismuth Subsalicylate/ 15ml of Suspension

Product Form: Suspension

Other Ingredients:
Benzoic Acid, Colors, Flavor, Magnesium Aluminum Silicate, Methylcellulose, Saccharin Sodium, Salicylic Acid, Sodium Salicylate, Sorbic Acid, Water.

PEPTO DIARRHEA CONTROL

Manufacturer: Procter & Gamble Company

Active Ingredient & Concentration:
1mg Loperamide Hydrochloride/5ml

Product Form:
Liquid

Other Ingredients:
Alcohol(5.25%), Citric Acid, Flavors, Glycerin, Methylparaben, Propylparaben, Purified Water.

RHEABAN, MAXIMUM STRENGTH

Manufacturer: Pfizer Consumer Health Care

Active Ingredient & Concentration:
750mg Activated Attapulgite/Tablet

Product Form:
Tablet

Other Ingredients:
Croscarmellose Sodium, Carnauba Wax, Hydroxypropylcellulose, Hydroxypropyl Methylcellulose 2910, Pectin, Pharmaceutical Glaze, Sucrose, Talc, Titanium Dioxide, Zinc Stearate, Colors.

PART II

SUPPLIER'S FORMULATIONS

KAOLIN-PECTIN SUSPENSION

Supplier: American Colloid Company

	Ingredient:	Wt.%
	Magnesium Aluminum Silicate (**Magnabrite HV**)	0.88
A	Carboxymethylcellulose	0.22
	Water	79.12
B	Kaolin	17.50
	Pectin	0.44
C	Saccharin	0.09
	Glycerin	1.75
	Flavor, Preservative	q.s.
		100.00

Mixing Procedure:

Dry blend **Magnabrite HV** and Carboxymethylcellulose. Slowly add the blend to water, agitating continually until smooth. Add Part B and continue stirring. Blend the Part C ingredients and add to Parts A + B. Mix until smooth.

CROSPOVIDONE ANTIDIARRHEAL EFFERVESCENT TABLET

Supplier: BASF Corp.

	Ingredient:	
	Crospovidone (**Crospovidone M**)	1000.0gm
A	Citric Acid	150.0gm
	Silicon Dioxide	25.0gm
	Sucrose	100.0gm
B	Saccharin Sodium	1.0gm
	Water	q.s.
	Sodium Bicarbonate	125.0gm
C	Flavor	65.0gm
	Magnesium Stearate	5.0gm

Mixing Procedure:

Granulate the Part A ingredients with the Part B solution, pass through a 0.8mm sieve, dry and mix with the Part C ingredients. Press with a rotary tablet press with medium compressive force to tablets weighing 1600mg with a diameter of 20mm.

BISMUTH CARBONATE SUSPENSION

Supplier: FMC Corp.

	Ingredient:	gm/100.0ml
A	Light Kaolin	26.66
B	Pectin	0.83
C	Bismuth Carbonate	0.67
D	Microcrystalline Cellulose(**Avicel® RC-591**)	0.94
E	Methylparaben	0.14
F	Saccharin Sodium	0.02
G	Aspartame	0.04
H	Sorbitol	4.00ml
I	Ethanol	0.50ml
J	Deionized water	q.s

Mixing Procedure:

Dissolve Part B in hot water. Disperse Part A in 75ml room temperature water. With constant agitation add Part C and continue stirring. Mix and cool to room temperature. Disperse **Avicel RC-591** in Water. Add it to the batch. Dissolve Part E in Part I and add it to the batch. Dissolve Parts F and G in Water and add. Add Flavor, Color, Sorbitol and Water to volume. Pass through homogenizer or colloid mill. Deaerate, if desired.

KAOLIN WITH PECTIN SUSPENSION

Supplier: FMC Corp.

Ingredient:	gm/100.0ml
Aspartame	0.04
Saccharin Sodium	0.02
Kaolin	20.00
Pectin	1.00
Glycerin	2.00ml
Methylparaben	0.12
Propylparaben	0.08
Carboxymethylcellulose Sodium (Medium Viscosity)	0.075
Microcrystalline Cellulose (**Avicel**^{R} **RC-591**)	0.75
Distilled water	q.s.
Sorbitol, 70% Solution	6.00ml

Mixing Procedure:

Disperse the **Avicel RC-591** and preservatives in 75 ml of distilled water for at least 30 minutes using moderate shear. Add ingredient Part H and mix until completely dispersed. Heat the **RC-591**/Carboxymethylcellulose Sodium dispersion to 75°C and slowly add the Pectin. Mix until dissolved (about 30 minutes). Cool above dispersion and add the glycerin while mixing. Disperse the Kaolin in the above mixture until homogeneous, with gentle mixing to avoid excess air entrapment. Dissolve Aspartame and Sodium Saccharin, in the Sorbitol solution and add to the batch. Adjust to volume and mix. Homogenize and/or deaerate if necessary.

KAOLIN AND PECTIN SUSPENSION

Supplier: FMC Corp.

Ingredient:	Wt.%
Kaolin	19.00
Pectin	0.40
Glycerin	2.00
Sodium Benzoate	0.05
Potassium Sorbate	0.05
Microcrystalline Cellulose (**Avicel RC-591**)	0.90
Xanthan Gum	0.10
Water	77.50
	100.00

Mixing Procedure:

Measure, by volume, 94% of the total amount of Water and slowly add **Avicel RC-591** powder while mixing. Agitate until the **Avicel** is fully dispersed. Slowly add Xanthan Gum while mixing. Mix until dissolved. Dry blend the Pectin with an equal quantity of Kaolin. Heat the **Avicel** dispersion to 70-80°C and add the Kaolin/Pectin blend. Mix until the pectin is dissolved, then remove from heat. As the dispersion cools, add the remaining Kaolin. Mix until homogeneous. Add Glycerin and mix until homogeneous. Mix the remaining Water, Sodium Benzoate and Potassium Sorbate and add to the suspension with mixing. Add Water to volume and mix until homogeneous.

KAOLIN-PECTIN SUSPENSION

Supplier: Southern Clay Products, Inc.

	Ingredient:	Wt.%
	Montmorillonite (**Gelwhite GP**)	1.00
A	Carboxymethylcellulose Sodium (Medium Visc.)	0.20
	Water	75.62
B	Kaolin USP	20.10
	Pectin	1.00
C	Saccharin	0.08
	Glycerin	2.00
D	Flavor, Preservative	q.s.
		100.00

Mixing Procedure:

Dry blend **Gelwhite GP** and CMC. Slowly add them to the water while agitating at maximum available shear and continue mixing until smooth. Add Part B ingredients slowly and continue mixing until smooth and uniform. Blend Part C ingredients and add to other components. Mix until smooth and uniform. Add flavor and preservative and mix until uniform

---

ANTIDIARRHEAL ORAL SUSPENSION

Supplier: R.T. Vanderbilt Company, Inc.

	Ingredient:	Wt.%
	Magnesium Aluminum Silicate (**Veegum K**)	1.20
A	Xanthan Gum (**Rhodigel**)	0.40
	Deionized Water	96.25
	Methylparaben	0.20
	Propylparaben	0.10
B	Saccharin Sodium	0.10
	Bismuth Subsalicylate, Powder	1.75
	Salicylic Acid to pH 4.0	q.s.
		100.00

Mixing Procedure:

Dry blend **Veegum K** and **Rhodigel** and add them to the water while mixing with maximum available shear. Continue mixing until smooth and uniform. Add Part B ingredients in the order listed, mixing each until uniformly dissolved or dispersed.

CHAPTER 5

ANTIEMETIC PRODUCTS

Antiemetic products are designed to control or prevent the symptoms of nausea and vomiting due to motion sickness, mild infectious diseases or pregnancy. The primary active ingredients used in the brand name medications are Dimenhydrinate, Cyclizine Hydrochloride and Meclizine Hydrochloride.

The products are available as tablets or liquids.

PART I

BRAND NAME FORMULATIONS

DRAMAMINE

Manufacturer: The Upjohn Company

Active Ingredient & Concentration:	Product Forms:
50mg Dimenhydrinate/Tablet	Tablet Tablet, Ch.

Other Ingredients:

Tablets contain Silicon Dioxide, Croscarmellose Sodium, Lactose, Microcrystalline Cellulose, Magnesium Stearate.

Chewable tablets contain Aspartame, Citric Acid, Colors, Flavor, Magnesium Stearate, Methacrylic Acid Copolymer, Sorbitol.

DRAMAMINE, CHILDREN'S

Manufacturer: The Upjohn Company

Active Ingredient & Concentration: 12.5mg Dimenhydrinate/5ml

Product Form: Liquid

Other Ingredients:
Alcohol (5%), FD&C Red No. 40, Flavor, Glycerin, Methylparaben, Sucrose, Water.

DRAMAMINE II

Manufacturer: The Upjohn Company

Active Ingredient & Concentration: 25mg Meclizine Hydrochloride/Tablet

Product Form: Tablet

Other Ingredients:
Colloidal Silicon Dioxide, Corn Starch, Lactose, D&C Yellow No. 10 Lake, Microcrystalline Cellulose, Magnesium Stearate.

EMETROL

Manufacturer: Bock Pharmacal Company

Active Ingredients & Concentration: 1.87g Dextrose + 1.87g Levulose + 21.5mg Phosphoric Acid/5ml

Product Form: Liquid

Other Ingredients:
Glycerin, Methylparaben, Purified Water, Flavors, D&C Yellow No. 10, (in Lemon-Mint Flavor), FD&C Red No. 40 (in Cherry Flavor).

MAREZINE

Manufacturer: Burroughs Wellcome Company

Active Ingredient & Concentration: 50mg Cyclizine Hydrochloride/Tablet

Product Form: Tablet

Other Ingredients:
Corn Starch, Potato Starch, Dextrin, Lactose, Magnesium Stearate.

PART II

SUPPLIER'S FORMULATIONS

MOTION SICKNESS TABLET (50mg/Tablet)

Supplier: FMC Corp.

	Ingredient:	Wt.%
A	Dimenhydrinate Powder	21.74
B	Microcrystalline Cellulose (**Avicel® PH-101**)	21.74
C	Lactose	54.65
D	Croscarmellose Sodium (**Ac-Di-Sol® SD-711**)	1.00
E	Fumed Silicon Dioxide	0.43
F	Stearic Acid	0.22
G	Magnesium Stearate	0.22
		100.00

Mixing Procedure:

Screen Parts A, F and G separately through 40 mesh sieve. Blend Parts A, B, D and E in a "V" Blender for 3 minutes. Add Part C into the blender and mix 17 minutes. Add Part F and blend 3 minutes. Add Part G to the blender and mix 5 minutes. Tablet on Manesty Express 20 using 3/8" flat bevel punches to a hardness of 6kg. Average tablet weight = 228mg.

ANTIEMETIC SUPPOSITORIES

Supplier: Gattefossé s.a.

Ingredient:	Wt.%
Metoclopramide	1.00
Hard Fat(**Suppocire AM**)	99.00
	100.00

Mixing Procedure:

Heat **Suppocire AM** to 50°C. Add Metoclorpramide to the excipient at 40°C, mixing with a homogenizer. When the batch cools to 33°C, pour into a mold previous chilled to 0 to -5°C. Remove from the mold after 3 minutes.

CHAPTER 6

ANTIFUNGAL & ANTIBACTERIAL PRODUCTS

Brand name antifungal products are designed to inhibit the growth and reduce the number of fungal cells. They are used to treat such common fungal infections as athlete's foot, jock itch, ringworm and vaginal yeast infections.

Antibacterial products, by definition, contain active ingredients that kill bacteria. They are used to prevent infection following skin abrasions, cuts, bruises and the like.

In September 1993, the United States Food & Drug Administration issued the Final Monograph on active ingredients in antifungal OTC products. It states, " The active ingredient of the product consists of any one of the following within the specified concentration established for each ingredient: Clioquinol 3%, Haloprogin 1%, Miconazole Nitrate 2%, Povidone-Iodine 10%, Tolnaftate 1%, Undecylenic Acid, Copper Undecylenate and Zinc Undecylenate may be used individually or in any ratio that provides a total undecylenate concentration of 10 to 25%"[1].

PART I

BRAND NAME FORMULATIONS

BACTINE, ANTISEPTIC & ANESTHETIC

Manufacturer: Miles, Inc

Active Ingredients & Concentration:	Product Form:
0.13% Benzalkonium Chloride 2.5% Lidocaine Hydrochloride	Liquid

Other Ingredients:
Edetate Disodium, Fragrances, Octoxynol 9, Propylene Glycol, Purified Water.

BACTINE, FIRST AID ANTIBIOTIC & ANESTHETIC

Manufacturer: Miles, Inc

Active Ingredients & Concentration
5000 units Polymyxin B Sulfate
+ 400 units Bacitracin
+ 5mg Neomycin Sulfate
+ 10mg Diperodon Hydrochloride/gm

Product Form:
Ointment

Other Ingredients:
Mineral Oil, White Petrolatum.

CAMPHO-PHENIQE, TRIPLE ANTIBIOTIC & PAIN RELIEVER

Manufacturer: Sterling Winthrop, Inc

Active Ingredients & Concentration:
500 units Bacitracin Zinc
+ 5mg Neomycin Sulfate
+ 5000 units Polymyxin B Sulfate
+ 40mg Lidocaine Hydrochloride/gm

Product Form:
Ointment

Other Ingredients:
White Petrolatum

CRUEX

Manufacturer: Fisons Consumer Health

Active Ingredients & Concentration:
Powder: 10% Calcium Undecylenate
Spray Powder: 19% Undecylenate as Undecylenic Acid and Zinc Undecylenate
Cream: 20% Undecylenate as Undecylenic Acid and Zinc Undecylenate

Product Forms:
Powder
Cream

Other Ingredients:
Powder contains Colloidal Silicon Dioxide, Fragrance, Isopropyl Myristate, Talc.
Spray powder contains Fragrance, Isobutane, Isopropyl Myristate, Menthol, Talc, Trolamine.
Cream contains Fragrance, Glycol Stearate SE, Lanolin, Methylparaben, PEG-8 Laurate, PEG-6 Stearate, Propylparaben, Purified Water, Sorbitol, Stearic Acid, Trolamine, White Petrolatum.

DESENEX

Manufacturer: Fisons Consumer Health

Active Ingredients & Concentration:
- Cream, Ointment, Powder and Spray Powder: 25% Undecylenate as Undecylenic Acid and Zinc Undecylenate
- Spray Liquid: 1% Tolnaftate
- Foam: 10% Undecylenic Acid

Product Forms:
- Cream
- Ointment
- Powder
- Liquid
- Foam

Other Ingredients:

Cream and ointment contain Fragrance, Glycol Stearate SE, Lanolin, Methylparaben, PEG-8 Laurate, PEG-6 Stearate, Propylparaben, Purified Water, Sorbitol, Stearic Acid, Trolamine, White Petrolatum.

Powder contains Fragrance, Talc.

Spray powder contains Fragrance, Isobutane, Isopropyl Myristate, Menthol, Talc, Trolamine.

Liquid contains BHT, Fragrance, Isobutane, Polyethylene Glycol 400, Alcohol SD 40B (41% w/w).

Foam contains Emulsifying Wax, Fragrance, Isobutane, Isopropyl Alcohol (29.2% w/w), Purified Water, Sodium Benzoate, Trolamine.

GYNE-LOTRIMIN

Manufacturer: Schering-Plough HealthCare Products

Active Ingredients & Concentration:
- Cream: 1.0% Clotrimazole
- Suppository: 100mg Clotrimazole/Unit

Product Forms:
- Cream
- Suppository

Other Ingredients:

Cream contains Benzyl Alcohol, Cetearyl Alcohol, Cetyl Esters Wax, Octyldodecanol, Polysorbate 60, Purified Water, Sorbitan Monostearate.

Suppositories contain Corn Starch, Lactose, Magnesium Stearate, Povidone.

LOTRIMIN AF

Manufacturer: Schering-Plough HealthCare Products

Active Ingredients & Concentration:
All products contain:
1.0% Clotrimazole

Product Forms:
Cream
Solution
Lotion

Other Ingredients:
Cream contains Benzyl Alcohol, Cetearyl Alcohol, Cetyl Esters Wax, Octyldodecanol, Polysorbate, Sorbitan Monostearate, Water.
Solution contains Polyethylene Glycol.
Lotion contains Benzyl Alcohol, Cetearyl Alcohol, Cetyl Esters Wax, Octyldodecanol, Polysorbate, Sodium Biphosphate, Sodium Phosphate Dibasic, Sorbitan Monostearate, Water.

MICATIN, (Athlete's Foot or Jock Itch)

Manufacturer: Ortho Pharmaceutical Corp.

Active Ingredients & Concentration:
All product forms contain:
2% Miconazole Nitrate

Product Forms:
Cream
Powder
Liquid

Other Ingredients:
Athlete's foot and jock itch creams contain Benzoic Acid, BHA, Mineral Oil, Peglicol 5 Oleate, Pegoxol 7 Stearate, Purified Water.
Athlete's foot powder contains Talc.
Athlete's foot and jock itch spray powders contain Alcohol, Propellant A-46, Talc, Stearalkonium Hectorite, Sorbitan Sesquioleate, Talc. May also contain Fragrance.
Athlete's foot antifungal spray liquid contains Alcohol, Benzyl Alcohol, Cocamide DEA, Propellant A-46, Sorbitan Sesquioleate, Tocopherol.

MONISTAT 7

Manufacturer: Ortho Pharmaceutical Corp.

Active Ingredients & Concentration:
Cream: 2.0% Miconazole Nitrate
Suppository: 100mg Miconazole Nitrate/Unit

Dosage Forms:
Cream
Suppository

Other Ingredients:
Cream contains Benzoic Acid, BHT, Mineral Oil, Peglicol 5 Oleate, Pegoxyl 7 Stearate, Purified Water.
Suppositories contain Hydrogenated Vegetable Oil Base.

MYCELEX-7

Manufacturer: Miles, Inc.

Active Ingredients & Concentration:
Cream: 1.0% Clotrimazole
Suppository: 100mg Clotrimazole/Unit

Product Forms:
Cream
Suppository

Other Ingredients:
Suppositories contain Corn Starch, Lactose, Magnesium Stearate, Povidone.
Cream contains Benzyl Alcohol, Cetostearyl Alcohol, Cetyl Esters Wax, Octyldodecanol, Polysorbate 60, Purified Water, Sorbitan Monostearate.

MYCELEX OTC

Manufacturer: Miles, Inc.

Active Ingredient & Concentration:
1.0% Clotrimazole

Product Form:
Liquid

Other Ingredients:
Polyethylene Glycol 400.

MYCITRACIN,MAXIMUM STRENGTH

Manufacturer: The Upjohn Company

Active Ingredients & Concentration:
500 units Bacitracin + Neomycin Sulfate equal to 3.5mg Neomycin + 5000 units Polymyxin B Sulfate/gm

Product Form:
Ointment

Other Ingredients:
Butylparaben, Cholesterol, Methylparaben, Mineral Oil, Microcrystalline Wax, Mineral Oil, White Petrolatum.

MYCITRACIN, PLUS PAIN RELIEVER

Manufacturer: The Upjohn Company

Active Ingredients & Concentration:
500 units Bacitracin + Neomycin Sulfate equal to 3.5mg Neomycin + 5000 units Polymyxin B Sulfate + 40mg Lidocaine/gm

Product Form:
Ointment

Other Ingredients:
Butylparaben, Cholesterol, Methylparaben, Mineral Oil, Microcrystalline Wax, White Petrolatum.

NEOSPORIN

Manufacturer: Burroughs Wellcome Company

Active Ingredients & Concentration:
5000 units Polymyxin B Sulfate + 400 units Bacitracin Zinc + 3.5mg Neomycin

Product Form:
Ointment

Other Ingredients:
White Petrolatum.

NEOSPORIN PLUS CREAM

Manufacturer: Burroughs Wellcome Company

Active Ingredients & Concentration:	Product Form:
10,000 units Polymyxin B Sulfate + 40mg Lidocaine + 3.5mg Neomycin/gm	Cream

Other Ingredients:
Methylparaben, Emulsifying Wax, Mineral Oil, Poloxamer 188, Propylene Glycol, Purified Water, White Petrolatum.

NEOSPORIN PLUS OINTMENT

Manufacturer: Burroughs Wellcome Company

Active Ingredients & Concentration:	Product Form:
10,000 units Polymyxin B Sulfate + 500 units Bacitracin Zinc + 40mg Lidocaine + 3.5mg Neomycin/gm	Ointment

Other Ingredients:
White Petrolatum.

POLYSPORIN

Manufacturer: Burroughs Wellcome Company

Active Ingredients & Concentration:	Product Forms:
10,000 units Polymyxin B Sulfate 500 units Bacitracin Zinc/gm of Ointment or Powder	Ointment Powder

Other Ingredients:
Ointment contains White Petrolatum.
Powder contains Lactose.

TINACTIN

Manufacturer: Schering-Plough HealthCare Products

Active Ingredients & Concentration:
All products forms contain:
1% Tolnaftate

Product Forms:
Cream
Liquid
Powder

Other Ingredients:
Antifungal cream contains BHT, Carbomer, Monoamylamine, PEG-8, Propylene Glycol, Titanium Dioxide.
Jock itch cream contains Cetearyl Alcohol, Ceteareth-30, Chlorocresol, Mineral Oil, Petrolatum, Propylene Glycol, Sodium Phosphate, Water.
Antifungal liquid contains BHT, Polyethylene Glycol.
Antifungal spray liquid contains Alcohol SD40-2 (36% w/w), BHT, PPG-12, Buteth-16.
Antifungal powder contains Corn Starch, Talc.
Antifungal and jock itch spray powders contain Alcohol SD40 or SD40-2 (14% w/w), BHT, Buteth-16, PPG-12, Hydrocarbon Propellant, Talc. (Deodorant powder also contains Starch/Acrylates/Acrylamide Copolymer, Fragrance).

TING

Manufacturer: Fisons Consumer Health

Active Ingredients & Concentration:
All product forms contain:
1% Tolnaftate

Product Forms:
Cream
Liquid
Powder

Other Ingredients:
Cream contains BHT, Fragrance, Polyethylene Glycol 400, Polyethylene Glycol 3350, Titanium Dioxide.
Powder contains Corn Starch, Fragrance, Talc.
Powder spray contains BHT, Fragrance, Isobutane, PPG-12-Buteth-16, Alcohol SD 40-B (14% w/w), Talc.
Liquid spray contains BHT, Fragrance, Isobutane, Polyethylene Glycol 400, Alcohol S 40-B (41% w/w).

PART II

SUPPLIER'S FORMULATIONS

METRONIDAZOLE VAGINAL EFFERVESCENT TABLET (500mg)

Supplier: BASF Corp.

	Ingredient:	Wt.%
	Metronidazole	27.93
A	Sodium Bicarbonate	33.52
	Povidone (**Kollidon 30**)	0.56
B	Povidone (**Kollidon 30**)	0.56
	Isopropanol or Ethanol, 96%	6.71
C	Tartaric Acid	27.93
	Polyethylene Glycol 6000	2.79
		100.00

Mixing Procedure:

Granulate Part A with the Part B solution and sieve. Mix with Part C and press into tablets of 1700mg weight and 16mm diameter.

POVIDONE-IODINE SEAMLESS SOLUTION

Supplier: BASF Corp.

Ingredient:	Wt.%
Povidone-Iodine (**PVP-Iodine 30/06**)	10.00
Hydroxyethylcellulose, High Viscosity	1.00
Sodium Hydroxide	3.15
Water	85.85
	100.00

Mixing Procedure:

Dissolve Povidone-Iodine in water. Slowly add Hydroxyethylcellulose to the mixture with stirring. After a clear solution is obtained, add the Sodium Hydroxide.

POVIDONE-IODINE OINTMENT

Supplier: BASF Corp.

	Ingredient:	Wt.%
A	Povidone-Iodine (**PVP-Iodine 30/06**)	10.00
B	Polyethylene Glycol 400	58.00
C	Polyethylene Glycol 4000	25.00
D	Sodium Bicarbonate	0.30
E	Water	6.70
		100.00

Mixing Procedure:

Dissolve Part D in the Part E. Add to this, 30gm of Part B. Slowly dissolve Part A, with stirring, in the above solution. Prepare a separate solution of the remaining Part B and Part C by heating to 60-70°C. Combine the two solutions at 50-60°C. Cool to room temperature with stirring.

POVIDONE-IODINE SOLUTION

Supplier: BASF Corp.

Ingredient:	
Povidone-Iodine (**PVP-Iodine 30/06**)	10.00gm
Citric Acid-Sodium Phosphate Buffer Soln.	90.00ml

Mixing Procedure:

Mix 48.5ml of 0.1 molar Citric Acid solution with 51.5ml of 0.2 molar Disodium Hydrogen Phosphate Dodecahydrate. To 90ml of the above buffer, slowly add the Povidone-Iodine with stirring until a clear, brown solution is obtained.

POVIDONE-IODINE SURGICAL SCRUB

Supplier: BASF Corp.

	Ingredient:	Wt.%
A	Povidone-Iodine (**PVP-Iodine 30/06**)	7.50
B	Ammonium Nonoxynol-4-Sulfate	24.00
C	Lauramide DEA	4.00
D	Water	64.50
		100.00

Mixing Procedure:

Dissolve Part C in 15ml of Part D at 60°C. Add the remaining Part D and dissolve the Part A in the solution with stirring. After a clear solution is obtained, add the Part B.

ANTI-INFECTIVE TOPICAL CREAM

Supplier: Gattefossé s.a.

	Ingredient:	Wt.%
	PEG-6 Stearate(and)Glyceryl Stearate (and)PEG-32 Stearate (**Tefose 63**)	20.00
	Mineral Oil	3.00
A	Apricot Kernel Oil PEG-6 Esters (**Labrafil M 1944 CS**)	3.00
	Demineralized Water	71.80
	Benzoic Acid	0.20
B	Isoconazole Nitrate	2.00
		100.00

Mixing Procedure:

Mix and heat Part A ingredients to 75°C. Allow to cool with gentle stirring. At 30°C, add Part B. Mix and cool to room temperature.

FOOT CARE MICROEMULSION

Supplier: Gattefossé s.a.

	Ingredient:	Wt.%
	Ethoxydiglycol (**Transcutol**)	15.50
A	Polyglyceryl-6 Dioleate (**Plurol Oleique**)	13.00
	PEG-8 Caprylic/Capric Glycerides (**Labrasol**)	45.00
B	Tolnaftate	1.00
	Demineralized Water	10.00
C	Apricot Kernel Oil PEG-6 Esters (**Labrifil M 1944 CS**)	5.00
	Caprylic/Capric Triglycerides (**Labrafac Lipophile**)	10.00
	Chlorocresol	0.50
		100.00

Mixing Procedure:

Mix the Part A ingredients and dissolve Part B in the mixture (1-2 hrs.). Add the Part C ingredients and mix until uniform.

FOOT CARE CREAM I

Supplier: Gattefossé s.a.

	Ingredient:	Wt.%
	Glyceryl Stearate(and)PEG-75 Stearate (**Gelot 64**)	15.00
	Hydrogenated Palm/Palm Kernel Oil PEG-6 Esters (**Labrafil M 2130 CS**)	2.00
A	Mineral Oil	6.00
	Sorbic Acid	0.05
	Sodium Methylparaben	0.05
	Demineralized Water	50.90
	Undecylenic Acid	5.00
B	Zinc Undecylenate	20.00
	Tolnaftate	1.00
		100.00

Mixing Procedure:

Mix and heat the Part A ingredients to 75°C. Allow to cool with gentle stirring. At 30 °C, add Part B. Homogenize, if necessary.

FOOT CARE CREAM II

Supplier: Gattefossé s.a.

	Ingredient:	Wt.%
	Glyceryl Stearate (**Gelol**)	5.00
	Propylene Glycol Stearate (**Monosterol**)	5.00
A	Octyldodecyl Myristate (**M.O.D.**)	10.00
	Isosearyl Isostearate	5.00
	Dimethicone, 100cs.	2.00
B	Demineralized Water	65.10
	Sucrose Distearate (**Sucroester 7**)	5.00
	Phenoxyethanol(and)Methylparaben(and) Ethylparaben(and)Propylparaben	0.40
C	Propylene Glycol	2.00
	Triclosan	0.30
D	Fragrance	0.20
		100.00

Mixing Procedure:

Heat the Part A and Part B ingredients, separately to 75°C. Add Part A to Part B with turbine mixing for one minute. Cool with gentle stirring. Add the Parts C and D ingredients, with mixing, at 30-35°C.

ABACTERIAL GEL

Supplier: BF Goodrich Co.

Ingredient:	Wt.%
Chlorhexidine	0.10
Demineralized Water	96.40
Carbomer 934P or 974P(**Carbopol 934P or 974P**)	1.00
Ethanol	2.40
Povidone	0.10
	100.00

Mixing Procedure:

Contact supplier for mixing procedure.

ANTIFUNGAL FOOT POWDER

Supplier: Inolex Chemical Company

Ingredient:	Wt.%
Dichlorobenzyl Alcohol (**Myacide SP**)	0.50
Allantoin	0.50
Corn Starch	20.00
Talc	79.00
	100.00

Mixing Procedure:
Carefully blend all ingredients and fill.

PROFESSIONAL CARE ANTIMICROBIAL SOAP

Supplier: NIPA Laboratories

Ingredient:	Wt.%
Sodium C14-16 Olefin Sulfonate	7.00
Hexylene Glycol	3.00
Chloroxylenol (**Nipacide PX**)	3.00
Cocamidopropyl Betaine	5.00
Water	82.00
	100.00

Mixing Procedure:
Dissolve Chloroxylenol in Hexylene Glycol. Mix Sodium C12-14 Olefin Sulfonate and Cocamidopropyl Betaine and add to above mixture. Add water and adjust the pH to 5.6-5.8.

ATHLETE'S FOOT CREAM

Supplier: R.T. Vanderbilt Company, Inc.

	Ingredient:	Wt.%
	Magnesium Aluminum Silicate (**Veegum**)	0.75
	Water	48.75
A	Sorbitol, 70%	10.00
	Polysorbate 80	1.00
	Zinc Undecylenate	20.00
	Caprylic Acid	5.00
	C12-15 Alcohols Benzoate	3.00
B	Polysorbate 80	1.50
	C18-36 Acid	2.00
	Glyceryl Stearate(and)PEG-100 Stearate	8.00
C	Preservatives	q.s.
		100.00

Mixing Procedure:

Slowly add **Veegum** to the Water, mixing with maximum available shear until smooth. Add the remaining Part A ingredients in order shown, mixing each until uniform. Avoid incorporating air, Heat Part A, with stirring, to 70-75°C. Heat Part B ingredients , with stirring, to 70-75°C. Add Part B to Part A and mix while cooling. Add Part C and fill at 45-50°C.

---

[1] *Federal Register,* 58, No. 183, Sept. 23, 1993, p49898

CHAPTER 7

ANTIPERSPIRANTS

The active ingredients in OTC antiperspirant products include aluminum salts and complexes thereof and aluminum zirconium salts and complexes. The most commonly used of these are Aluminum Chlorohydrate and Aluminum Zirconium Tetrachlorohydrex GLY. Because it is not customary to indicate the concentration of the active ingredient(s) on the product label, Part I of this chapter lists only the active ingredient but not its concentration in the product.

The Tentative Final Monograph for OTC antiperspirants was published by the United States Food and Drug Administration in the *Federal Register* on August 20, 1982 and includes the Category I antiperspirant active ingredients and the maximum allowable concentration of each.

The brand name products are marketed in as many as four different forms; stick, roll-on, cream and/or aerosol spray.

PART I

BRAND NAME FORMULATIONS

ALMAY, SUPERSENSITIVE

Manufacturer: Almay, Inc.

Active Ingredients: Product Form:
Regular Stick: Aluminum Chlorohydrate — Stick
Wide Solid Stick: Aluminum Zirconium Tetrachlorohydrex GLY

Other Ingredients:
Cyclomethicone, Stearyl Alcohol, Hydrogenated Castor Oil, Talc, PEG-8 Distearate, Silicon Dioxide, Propylparaben, Titanium Dioxide. Wide solid stick also contains Bisabolol.

ALMAY, ROLL-ON

Manufacturer: Almay, Inc.

Active Ingredient:
Aluminum Zirconium
Tetrachlorohydrex GLY

Product Form:
Roll-On

Other Ingredients:
Water, Glyceryl Stearate, Laureth-23, Magnesium Aluminum Silicate, Laureth-4, Lapyrium Oil, Lauric Acid, EDTA.

ALMAY, ALL DAY PROTECTION

Manufacturer: Almay, Inc.

Active Ingredient:
Aluminum Chlorohydrate

Product Form:
Roll-On

Other Ingredients:
Cyclomethicone, Quaternium-18 Hectorite, Alcohol, SD40B, Corn Starch Polysaccharide, Microcrystalline Cellulose, Fragrance, Water.

ARRID, EXTRA DRY

Manufacturer: Carter-Wallace, Inc.

Active Ingredient:
Stick: Aluminum Zirconium
Tetrachlorohydrex GLY
Aerosol: Aluminum Chlorohydrate

Product Forms:
Stick
Aerosol

Other Ingredients:
Stick contains Cyclomethicone, Dipropylene Glycol, Dimethicone Copolyol, Laureth-23, Mineral Oil, Dimethicone, PEG-8 Laurate, Fragrance, Lanolin Alcohol, Water.
Aerosol contains Isobutane, Propane, Isopropyl Palmitate, Cyclomethicone, Dibutyl Phthalate, Quaternium-18 Bentonite, Fragrance.

ARRID XX

Manufacturer: Carter-Wallace, Inc.

Active Ingredient:
Both forms contain: Aluminum Zirconium Tetrachlorohydrex GLY

Product Forms:
Stick
Aerosol

Other Ingredients:
Stick contains Cyclomethicone, Fragrance, Diisopropyl Adipate, PPG-14 Butyl Ether, Stearyl Alcohol, Talc, Hydrogenated Castor Oil, Laureth-4, Colors.
Aerosol contains Isobutane, Propane, Isopropyl Palmitate, Cyclomethicone, Dibutyl Phthalate, Quaternium-18 Bentonite, Fragrance.

ARRID, TEEN IMAGE

Manufacturer: Carter-Wallace, Inc.

Active Ingredient:
Aluminum Zirconium Tetrachlorohydrex GLY

Product Form:
Stick

Other Ingredients:
Cyclomethicone, Fragrance, Stearyl Alcohol, Talc, Hydrogenated Castor Oil, Colors, PEG-8 Distearate, Silicon Dioxide, Titanium Dioxide, Propylparaben.

BAN

Manufacturer: Bristol-Myers Products

Active Ingredients:
Roll-On: Aluminum Chlorohydrate
Stick: Aluminum Zirconium Tetrachlorohydrex GLY

Product Forms:
Roll-On
Stick

Other Ingredients:
Roll-On contains Water, Fragrance, Colors, PPG-11 Stearyl Ether, Steareth-2, Steareth-20, Rose Oil, Mandarin Orange Oil, Chamomile Oil, Edetate Disodium.
Stick contains Cyclomethicone, Fragrance, Hydrogenated Castor Oil, Stearyl Alcohol, Talc, BHT, Hydrogenated Polyisobutene, PEG-25 Propylene Glycol Stearate, Glyceryl Stearate.

BAN, CLEAR STICK

Manufacturer: Bristol-Myers Products

Active Ingredient:
Aluminum Zirconium Tetrachlorohydrex GLY

Product Form:
Stick

Other Ingredients:
Fragrance, Propylene Glycol, FD&C Blue No. 1, Glycerin, Poloxamer 217, Sodium Hydroxide, Glycereth-7 Benzoate, Dipropylene Glycol, Dibenzylidene Sorbitol, Diisopropyl Sebacate.

BAN, FOR MEN

Manufacturer: Bristol-Myers Products

Active Ingredient:
Aluminum Zirconium Tetrachlorohydrex GLY

Product Form:
Stick

Other Ingredients:
Fragrance, Propylene Glycol, FD&C Blue No. 1, Glycerin, Poloxamer 217, Sodium Hydroxide, Glycereth-7 Benzoate, Dipropylene Glycol, Dibenzylidene Sorbitol, Diisopropyl Sebacate.

DEGREE

Manufacturer: Helene Curtis Industries

Active Ingredients:
Aerosol: Aluminum Chlorohydrate
Stick and Roll-On: Aluminum Zirconium Tetrachlorohydrex GLY

Product Forms:
Aerosol
Stick
Roll-On

Other Ingredients:
Aerosol contains Cyclomethicone, Fragrance, Isobutane, propylene carbonate, Starch, Dextrin, C12-15 Alcohols Benzoate, Quaternium-18 Hectorite, Hexamethyldisiloxane.
Stick contains Cyclomethicone, Fragrance, Hydrogenated Castor Oil, Starch, Stearyl Alcohol, Talc, C12-15 Alcohols Benzoate, Arachidyl Alcohol.
Roll-On contains Cyclomethicone, Fragrance, Propylene Carbonate, Silicon Dioxide, Starch, Aluminum Starch Octenyl Succinate, Quaternium-18 Hectorite, Dioctyl Adipate.

GILLETTE

Manufacturer: The Gillette Company

Active Ingredient:
Aluminum Zirconium Tetrachlorohydrex GLY

Product Form:
Stick

Other Ingredients:
Cyclomethicone, Stearyl Alcohol, PPG-10 Butanediol, C12-15 Alcohols Benzoate, Hydrogenated Castor Oil, Fragrance/Starch/Polyol, Myristyl Myristate, PEG-8 Distearate, Fragrance.

LADY MITCHUM

Manufacturer: Revlon Consumer Products, Inc.

Active Ingredient:
Aluminum Zirconium Tetrachlorohydrex GLY

Product Form:
Stick

Other Ingredients:
Cyclomethicone, Fragrance, Hydrogenated Castor Oil, Silicon Dioxide, Stearyl Alcohol, Talc, PEG-8 Distearate, Propylparaben, Talc, Titanium Dioxide.

LADY SPEED STICK

Manufacturer: The Mennen Company

Active Ingredient:
Aluminum Zirconium Tetrachlorohydrex GLY

Product Form:
Stick

Other Ingredients:
Cyclomethicone, Stearyl Alcohol, Glyceryl Stearate, PEG-100 Stearate, Fragrance, Talc, Hydrogenated Castor Oil, PPG-14 Butyl Ether, Allantoin, Colors.

MITCHUM

Manufacturer: Revlon Consumer Products, Inc.

Active Ingredient:
Both forms: Aluminum Zirconium Tetrachlorohydrex GLY

Product Forms:
Stick
Roll-On

Other Ingredients:
Stick contains Cyclomethicone, Hydrogenated Castor Oil, Silica, Stearyl Alcohol, Talc, Titanium Dioxide, PEG-8 Distearate, Propylparaben. May also contain Fragrance.
Roll-On contains Water, EDTA, Lapyrium Chloride, Lauric Acid, Glyceryl Stearate, Laureth-23, Magnesium Aluminum Silicate, Laureth-4. May also contain Fragrance.

MITCHUM, SUPER DRY

Manufacturer: Revlon Consumer Products, Inc.

Active Ingredient:
Aluminum Zirconium Tetrachlorohydrex GLY

Product Form:
Roll-On

Other Ingredients:
Cyclomethicone, Dimethicone, Silicon Dioxide, Alcohol SD40B, Quaternium-18 Hectorite, Pentadecalactone.

RIGHT GUARD, SPORT

Manufacturer: The Gillette Company

Active Ingredients:
Aerosol: Aluminum Chlorohydrate
Stick: Aluminum Zirconium Tetrachlorohydrex GLY

Product Forms:
Aerosol
Stick

Other Ingredients:
Aerosol contains Cyclomethicone, Fragrance, Isobutane, Propane, Butane, Isopropyl Myristate, Alcohol SD40. Quaternium-18 Hectorite.
Stick contains Cyclomethicone, Hydrogenated Castor Oil, Silica, Stearyl Alcohol, Fragrance/Starch/Polyol, Talc, PEG-8 Distearate, Fragrance.

SECRET

Manufacturer: The Procter & Gamble Company

Active Ingredients:
Stick, Roll-On and Cream: Aluminum Zirconium Tetrachlorohydrex GLY
Aerosol: Aluminum Chlorohydrate

Product Forms:
Stick
Roll-On
Cream
Aerosol

Other Ingredients:
Stick contains Cyclomethicone, Dimethicone, Hydrogenated Castor Oil, Propylene Glycol, Silicon Dioxide, Stearyl Alcohol, Talc, Dipropylene Glycol, Eicosanol or Behenyl Alcohol. May also contain Pentadecalactone and/or Fragrance.
Roll-On contains Cyclomethicone, Dimethicone, Polyethylene, Propylene Carbonate, Polydecene, Quaternium-18 Hectorite.
Aerosol contains Cyclomethicone, Dimethicone, Isobutane, Propane, Propylene Glycol, Isopropyl Myristate. May also contain Fragrance.
Cream contains Water, Isopropyl Palmitate, D&C Blue No. 1 Aluminum Lake, PEG-50 Stearate, Glyceryl Stearate, PEG-2 Stearate, Glycerin, Paraffin. May also contain Fragrance.

SHOWER TO SHOWER

Manufacturer: Johnson & Johnson Consumer Products, Inc

Active Ingredient:
Both forms: Aluminum Zirconium Tetrachlorohydrex GLY

Product Forms:
Stick
Roll-On

Other Ingredients:
Stick contains Cyclomethicone, Corn Starch, Fragrance, Hydrogenated Castor Oil, Isosorbite Laurate, PPG-14 Butyl Ether, Stearyl Alcohol, Talc.
Roll-On contains Cyclomethicone, Corn Starch, Fragrance, Dimethicone, Polyethylene, Quaternium-18 Hectorite/Propylene Carbonate, Silicon Dioxide.

SPEED STICK, SUPER DRY

Manufacturer: The Mennen Company

Active Ingredient:
Aluminum Zirconium Tetrachlorohydrex GLY

Product Form:
Stick

Other Ingredients:
Cyclomethicone, Stearyl Alcohol, Glyceryl Stearate, PEG-100 Stearate, Fragrance.

SPEED STICK, SENSITIVE SKIN

Manufacturer: The Mennen Company

Active Ingredient:
Aluminum Chlorohydrate

Product Form:
Stick

Other Ingredients:
Cyclomethicone, Aloe Extract, Stearyl Alcohol, Glyceryl Stearate, PEG-100 Stearate, Fragrance.

SURE

Manufacturer: The Procter & Gamble Company

Active Ingredients:
Stick and Roll-On: Aluminum Zirconium Tetrachlorohydrex GLY
Aerosol: Aluminum Chlorohydrate

Product Forms:
Stick
Roll-On
Aerosol

Other Ingredients:
Stick contains Cyclomethicone, Dimethicone, Hydrogenated Castor Oil, Polyethylene, Silicon Dioxide, Stearyl Alcohol, Talc, Dipropylene Glycol, Behenyl Alcohol. May also contain Fragrance and Pentadecalactone.
Roll-On contains Cyclomethicone, Dimethicone, Fragrance, Polyethylene, Propylene Carbonate, Polydecene, Quaternium-18 Hectorite.
Aerosol contains Cyclomethicone, Dimethicone, Propane, Propylene Carbonate, Ethylene Brassylate, Quaternium-18 Hectorite, Butane, Isopropyl Myristate. May also contain Fragrance,

---

PART II

SUPPLIER'S FORMULATIONS

ANTIPERSPIRANT STICK

Supplier: Amerchol Corp.

	Ingredient:	Wt.%
	Stearyl Alcohol (**Stearal**)	23.00
	Isopropyl Palmitate (**Propal**)	20.00
A	PPG-14 Butyl Ether (**UCON Fluid AP**)	10.00
	Hydrogenated Castor Oil	4.00
	Laureth-4 (**Ameroxol LE-4**)	1.00
	Talc	4.00
B	Aluminum Chlorohydrate, 100%	20.00
C	Cyclomethicone	18.00
	Fragrance	q.s.
		100.00

Mixing Procedure:

Mix the Part A ingredients and gently heat to 55°C. When uniform, Add Part B, maintaining temperature. Heat the Cyclomethicone to 55°C, add with stirring. Continue stirring and heat to 70°C. Cool and stir to 50°C, add Fragrance, pour into molds.

ECONOMICAL ANTIPERSPIRANT STICK

Supplier: Croda, Inc.

	Ingredient:	Wt.%
	PPG-2 Myristyl Ether Propionate (**Crodamol PMP**)	15.50
A	Synthetic Beeswax (**Syncrowax BB4**)	1.00
	Cetyl Alcohol (**Crodacol C-70**)	12.00
	Behenamide MEA (**Incromide BEM**)	12.00
	Cyclomethicone	37.00
	Aluminum Zirconium Tetrachlorohydrex GLY	20.00
B	Talc	1.00
	Fumed Silica	1.50
		100.00

Mixing Procedure:

Heat the Part A ingredients to 75°C omitting the Cyclomethicone. Premix Part B ingredients and add to Part A at 75 °C. Mix well. Continue stirring and add Cyclomethicone at 70°C. Fill at approximately 55-60°C.

SUSPENSION ANTIPERSPIRANT STICK

Supplier: Dow Corning Corp.

	Ingredient:	Wt.%
	Cyclomethicone (**DC 345 Fluid**)	55.00
A	Stearyl Alcohol	20.00
	PPG-14 Butyl Ether	2.00
	Hydrogenated Castor Oil	1.00
B	Talc	2.00
	Aluminum Zirconium Tetrachlorohydrex GLY	20.00
		100.00

Mixing Procedure:

Mix the Part A ingredients in a vessel equipped for reflux. Mix while heating to 70°C. Hold until all solids are dissolved. Add Part B ingredients, mix to completely disperse the solids. Mix while cooling to 53°C, then pour into suitable containers.

TRANSPARENT ANTIPERSPIRANT GEL

Supplier: Dow Corning Corp.

	Ingredient:	Wt.%
A	Cyclomethicone(and)Dimethicone Copolyol (**Dow Corning 3225C Formulation Aid**)	10.00
	Cyclomethicone (**Dow Corning 244 Fluid**)	7.00
B	Aluminum Chlorohydrate(and)Water	50.00
	Propylene Glycol	16.00
	Water	17.00
		100.00

Mixing Procedure:

Mix the Part A ingredients and measure the refractive index, Separately mix Part B ingredients and measure the refractive index. Adjust the refractive index of Part B to match Part A using Propylene Glycol to raise it or Water to lower it. When the indices match, mix Part A with a homogenizer and slowly add Part B. Mix until uniform.

ANTIPERSPIRANT GEL

Supplier: General Electric Company

	Ingredient:	Wt.%
A	Cyclomethicone (**SF1228**)	10.00
	Cyclomethicone (**SF1204**)	14.00
B	Polysorbate 80	0.25
	Aluminum Zirconium Tetrachlorohydrex GLY	20.00
	Water	55.75
		100.00

Mixing Procedure:

Mix the Part A ingredients, Dissolve Polysorbate 80 in warm Water. Add Aluminum Zirconium Tetrachlorohydrex GLY to the Water/Polysorbate 80 solution and mix. Add Part B to Part A with high shear mixing such as a homogenizer until uniform.

ANTIPERSPIRANT SOLID

Supplier: General Electric Company

Ingredient:	Wt.%
Cyclomethicone (**SF1202**)	45.00
Dimethicone (**SF-96-100**)	5.00
Stearyl Alcohol	19.00
Hydrogenated Castor Oil	3.00
Talc	4.00
Glyceryl Stearate(and)PEG-100 Stearate	2.00
Aluminum Zirconium Tetrachlorohydrex GLY	22.00
	100.00

Mixing Procedure:

Mix **SF1202** with Stearyl Alcohol. Add Aluminum Zirconium Tetrachlorohydrex GLY, Talc and Glyceryl Stearate(and)PPG-100 Stearate. Heat to 75°C and stir with moderate agitation until all the wax is melted. Melt the Hydrogenated Castor Oil and add to the above mixture as a liquid and stir for 15 minutes. Cool the mixture to 55°C with continual stirring and pour into containers. Avoid air entrapment.

ANTIPERSPIRANT SUSPENSION ROLL-ON

Supplier: General Electric Company

Ingredient:	Wt.%
Cyclomethicone (**SF1173**)	45.70
Cyclomethicone (**SF1202**)	19.50
Dimethicone (**SF-96-50**)	5.00
Quaternium-18 Hectorite	2.50
Ethanol	2.00
Aluminum Zirconium Tetrachlorohydrex GLY	25.00
Fumed Silica	0.30
	100.00

Mixing Procedure:

Mix **SF1173, SF1202** and Quaternium-18 Hectorite in a high speed mixer. Add **SF-96-50** and Ethanol and continue mixing. Add Fumed Silica and Aluminum Zirconium Tetrachlorohydrex GLY and mix for 15 minutes. Transfer to a homogenizer and mix 3 minutes at high speed.

ANTIPERSPIRANT STICK

Supplier: Henkel Corp.

	Ingredient:	Wt.%
	Aluminum Chlorohydrex PG	20.00
A	Fumed Silica	1.00
	Propylene Glycol	30.00
	Pentaerythrityl Tetrapelargonate (**Emerest 2486**)	15.00
B	Stearamide MEA	15.00
	Deionized Water	10.40
	Polysorbate 80(and)Cetyl Acetate(and)	
C	Acetylated Lanolin Alcohol (**Ethoxyol 1707**)	8.00
	Ethylene Bassylate (**Emeressence 1150**)	0.60
		100.00

Mixing Procedure:

Combine the Part A ingredients with high speed stirring. Continue stirring and heat to 80°C. Slowly add Stearamide MEA and stir until dissolved. Heat the Water to 80°C and add slowly to the mixture. Continue stirring for 30 minutes. Add the Part C ingredients, stirring until homogeneous. Pour into sticks at 60°C.

ANTIPERSPIRANT STICK WITH PROTECTANT PROTEIN

Supplier: Hormel & Company

	Ingredient:	Wt.%
	Cyclomethicone	47.00
A	Sodium Stearate	7.00
	Stearyl Alcohol	17.00
B	Propyltrimonium Hydrolyzed Collagen (**Protectein**)	5.00
C	Aluminum Zirconium Chlorohydrate	20.00
	Dimethicone	2.00
D	Propylene Glycol(and)diazolidinyl Urea(and) Methylparaben(and)Propylparaben	1.00
	Fragrance	1.00
		100.00

Mixing Procedure:

Combine the Part A ingredients in a suitable vessel and heat to 80°C with mixing. Slowly add Part B. Mix until uniform. Add Part C and mix until homogeneous. Add the Part D ingredients and mix well. At 50°C, pour into suitable molds and cool.

ANTIPERSPIRANT STICK

Supplier: Hüls America

	Ingredient:	Wt.%
	PEG-6 Caprylic/Capric Glycerides (**Softigen 767**)	40.00
	Sodium Stearate	8.00
A	Glycerin	10.00
	Sugar	8.00
	Water	28.00
	Preservative	q.s.
	Ethanol, 96%	3.30
B	Ethoxydiglycol(and)Diisopropyl Adipate(and) Methylparaben(and)Dehydroacetic Acid(and) Propylparaben(and)Sorbic Acid	0.70
	Fragrance	2.00
		100.00

Mixing Procedure:

Mix and heat Part A ingredients to 70°C. Add Part B ingredients to Part A. Mix and pour into molds.

AEROSOL ANTIPERSPIRANT

Supplier: ICI Americas, Inc.

	Ingredient:	Wt.%
	Quaternium-18 Hectorite Gel	8.00
A	PPG-15 Stearyl Ether (**Arlamol E**)	3.00
	Cyclomethicone	3.00
	Sorbitan Oleate (**Arlacel 80**)	0.50
B	Aluminum Chlorohydrate	6.00
C	Hydrocarbon Repellant A-46	79.50
		100.00

Mixing Procedure:

Mix the Part A ingredients thoroughly using medium shear equipment. Add Part B at a slow, steady rate and mix thoroughly. Fill, vacuum crimp and charge aerosol units with Part C.

AL ZR TETRACHLOROHYDREX-GLY STICK

Supplier: ICI Americas, Inc.

	Ingredient:	Wt.%
A	Cyclomethicone	48.50
	Isosorbide Laurate (**Arlamol ISML**)	10.00
B	Aluminum Zirconium Tetrachlorohydrex GLY	20.00
	Stearyl Alcohol	16.00
	Hydrogenated Castor Oil	5.00
C	Fragrance	0.50
		100.00

Mixing Procedure:

Stir and heat the Part B ingredients to 85°C or until the Hydrogenated Castor Oil is melted. Stir until the temperature drops to 60°. Add Part A to Part B at 60°C, stirring well. Add Part C and then pour into molds just above the set point.

ANTIPERSPIRANT ROLL-ON

Supplier: ICI Americas, Inc.

	Ingredient:	Wt.%
A	PPG-15 Stearyl Ether (**Arlamol E**)	3.00
	Isosorbide Laurate (**Arlamol ISML**)	4.00
	Steareth-21 (**Brij 721**)	1.50
	Steareth-2 (**Brij 72**)	2.50
B	Deionized Water	49.00
C	Aluminum Zirconium Tetrachlorohydrex GLY, 50% Solution	40.00
D	Preservative	q.s.
		100.00

Mixing Procedure:

Heat the Part A ingredients to 70°C and Part B to 72°C. Add Part B to Part A using a propeller stirrer. Add Part C at about 50°C and Part D at 45°C. At 35°C, replace water lost by evaporation and package.

CLEAR ROLL-ON ANTIPERSPIRANT LOTION

Supplier: ICI Americas, Inc.

	Ingredient:	Wt.%
A	Ethanol SDA-40	69.00
	Hydroxypropyl Cellulose	1.00
B	PPG-15 Stearyl Ether (**Arlamol E**)	10.00
C	Aluminum Clorhydroxide-Propylene Glycol Complex	20.00
		100.00

Mixing Procedure:

Mix the Part A ingredients thoroughly using high shear, if necessary. Add Part B and stir until homogeneous. Add Part C and stir until clear.

OIL-IN-WATER ANTIPERSPIRANT CREAM

Supplier: ICI Americas, Inc.

	Ingredient:	Wt.%
A	Steareth-21 (**Brij 721**)	5.00
	Stearyl Alcohol	6.00
B	Water	49.00
	Aluminum Zirconium Trichlorohydrate	40.00
		100.00

Mixing Procedure:

Heat the Part A ingredients to 70°C and the Part B ingredients to 72°C. Add Part B to Part A with agitation. Stir until the cream has set.

ROLL-ON ANTIPERSPIRANT LOTION

Supplier: ICI Americas, Inc.

	Ingredient:	Wt.%
	Cyclomethicone	10.00
A	Steareth-2 (**Brij 72**)	2.00
	Steareth-21 (**Brij 721**)	0.70
B	Water	52.30
C	Preservative	q.s.
D	Aluminum Chlorohydrate, 50% Solution	35.00
		100.00

Mixing Procedure:

Heat the Part A ingredients to 65-70°C and Part B several degrees higher. Add Part B to Part A with agitation. Cool to 35-40°C with stirring and add Parts C and D. Stir until homogeneous and package.

ANTIPERSPIRANT SPRAY

Supplier: Inolex Chemical Company

Ingredient:	Wt.%
Aluminum Chlorohydrate, 50%	40.00
Propylene Glycol	10.00
Hydrolyzed Collagen (**Lexein X250**)	3.00
2-Bromo-2-Nitropropane-1,3-Diol	0.05
Water	46.95
	100.00

Mixing Procedure:

Blend the ingredients at room temperature until clear. Fill into a lined aerosol container with A-46 propellant at an 80:20, concentrate:propellant ratio.

DRY ANTIPERSPIRANT STICK

Supplier: Inolex Chemical Company

	Ingredient:	Wt.%
A	Glyceryl Stearate(and)PEG-100 Stearate (**Lexemul 561**)	1.00
	Stearyl Alcohol	20.00
	PEG-20	5.00
B	Cyclomethicone	51.50
	Aluminum Chlorohydrate Powder	20.00
C	Talc	1.00
	Fumed Silica	1.50
		100.00

Mixing Procedure:

Melt the Part A ingredients together. Quickly add Part B and Part C ingredients to Part A in succession. Mix until homogeneous, pour into molds and cool.

STAY-DRY DEODORANT/ANTIPERSPIRANT STICK

Supplier: International Sourcing, Inc.

	Ingredient:	Wt.%
	Magnesium Aluminum Silicate	1.00
A	Deionized Water	20.00
	Aluminum Chlorohydrate, Impalpable	18.00
	Cyclomethicone	25.00
	Isopropyl Myristate	5.00
B	Stearyl Alcohol	24.00
	Sorbitan Trioleate (**Arlacel 85**)	5.00
	Polysorbate 85 (**Tween 85**)	1.00
C	**Tea Tree Oil**	1.00
		100.00

Mixing Procedure:

Slowly add Magnesium Aluminum Silicate to water while agitating with maximum available shear. Continue mixing until smooth. Add Aluminum Chlorohydrate and mix until smooth. Heat Part A, with stirring, to 70°C. Heat Part B, with stirring, to 65°C. Add Part A to Part B. Cool to 50°C and add Part C. Stir until uniform. Pour into molds and cool.

ANTIPERSPIRANT ROLL-ON

Supplier: Lonza, Inc.

	Ingredient:	Wt.%
	Magnesium Aluminum Silicate	1.00
	Hydrogenated Starch Hydrolyzate(**Hystar 7000**)	2.00
A	Water	48.00
	DMDM Hydantoin(and)Iodopropynyl Butyl Carbamate (**Glydant Plus**)	q.s.
	Myristyl Propionate (**Lonzest 143-S**)	3.00
	Cetyl Alcohol	2.00
B	PEG-98 Lanolin	2.00
	PEG-40 Stearate (**Pegosperse 1750MS**)	6.00
	Aluminum Chlorohydrate, 50% Solution	36.00
		100.00

Mixing Procedure:

Heat the Part A and the first four Part B ingredients separately to 75-80°C. Add Part A to Part B with agitation. Begin cooling. At 40°C, add Aluminum Chlorohydrate. Continue mixing while cooling.

ANTIPERSPIRANT STICK

Supplier: PPG Industries, Inc.

	Ingredient:	Wt.%
	Stearyl Alcohol	22.00
	Hydrogenated Castor Oil	1.00
A	Glyceryl Stearate(and)PEG-100 Stearate (**Mazol 165C**)	1.00
	Tetrabutoxypropyl Trimethylsiloxane (**Masil 756**)	5.00
B	Aluminum Zirconium Tetrachlorohydrex GLY	20.00
	Talc	2.00
C	Cyclomethicone (**Masil SF-V**)	49.00
	Fragrance	q.s.
		100.00

Mixing Procedure:

Blend and heat the Part A ingredients to 65°C. When molten and uniform, add Part B with good agitation to disperse them. Add the Cyclomethicone, maintaining the temperature at 60-65°C. When uniform, cool to 55°C add fragrance and fill.

ANTIPERSPIRANT LOTION

Supplier: R.I.T.A Corp.

Ingredient:	Wt.%
Stearyl Alcohol(and)Ceteareth-20 (**Ritapro 200**)	4.60
PEG-75 Lanolin (**Laneto 50**)	8.80
Mineral Oil, 65/75 Viscosity	9.00
Aluminum Chlorohydrate, (**Ritachlor, 50%**)	40.00
Distilled Water	37.60
	100.00

Mixing Procedure:

Melt **Ritapro 200** and Mineral Oil at 75°C. Add preheated water and **Laneto 50** and mix to form an emulsion. When cool, add **Ritachlor 50%** slowly. Continue mixing until uniform.

CLEAR ROLL-ON ANTIPERSPIRANT

Supplier: R.I.T.A Corp.

	Ingredient:	Wt.%
A	Hydroxyethylcellulose	0.20
	Distilled Water	29.70
	Benzethonium Chloride	0.10
B	Distilled Water	8.00
	Alcohol SD40	20.00
	PEG-75 Lanolin (**Laneto 50**)	7.00
C	Aluminum Chlorohydrate (**Ritachlor 50%**)	35.00
		100.00

Mixing Procedure:

Prepare Part A in advance to allow the gum to hydrate fully before incorporation into the formula. Slowly add Part B to Part A at room temperature with high speed propeller agitation. Slowly add Part C with high speed stirring. Fragrance, if any, should be added to Part B.

ANTIPERSPIRANT CREAM

Supplier: R.I.T.A Corp.

	Ingredient:	Wt.%
A	Zinc Ricinoleate(and)Propylene Glycol(and) Disodium PEG-8 Ricinolsuccinate(and) PEG-7 Glyceryl Cocoate(and)Glycerin(and) Triethanolamine (**Grillocin AT Basis)**	3.00
	PEG-5 Glyceryl Stearate	5.00
	Cetyl Alcohol (**RITA CA**)	3.00
	Caprylic/Capric Triglyceride	10.00
B	Aluminum Chlorohydrate (**Ritachlor 50%**)	12.00
	Glycerin	2.00
	Distilled Water	q.s.
C	Fragrance	q.s.
	Citric Acid to pH 4	q.s.
		100.00

Mixing Procedure:

Heat Part A and Part B separately to 70°C. Add the Part B to Part A and homogenize. Cool slowly to 45°C. Re-mix, add Fragrance and adjust th pH.

ENHANCED EFFICACY CREAM

Supplier: Reheis, Inc.

	Ingredient:	Wt.%
A.	Aluminum Chlorohydrate Solution (**Reach 501**)	44.50
B.	Glyceryl Stearate(and)PEG-100 Stearate	15.00
C.	Cetyl Alcohol	5.00
D.	Sorbitol, 70% Solution	3.00
E.	Deionized Water	32.50
F.	Fragrance	q.s.
		100.00

Mixing Procedure:

Combine Part B and Part C and heat to 75°C. Combine Part D and Part E and heat to 75°C. Add the water phase slowly to the Part B + C combination and cool to 55°C while agitating with an overhead mixer. Add Part A and mix thoroughly. Add Part F and cool to 35°C. Homogenize at 3000 psi.

ENHANCED EFFICACY HYDRO-ALCOHOLIC ROLL-ON

Supplier: Reheis, Inc.

	Ingredient:	Wt.%
A.	Aluminum Chlorohydrate Solution (**Reach 501**)	44.50
B.	PPG-5 Ceteareth-20	2.00
C.	Hydroxyethylcellulose	0.20
D.	Deionized Water	11.20
E	Alcohol SD40	42.10
F.	Fragrance	q.s.
		100.00

Mixing Procedure:

Disperse Part C in Part D. Mix until clear, about 2 hours. Add Part A gradually. Mix rapidly using overhead stirring to dissolve. In a separate container, combine Part B and Part E, then add slowly with constant agitation to the batch. Add Fragrance, mix thoroughly and pour into roll-on containers.

ENHANCED EFFICACY PUMP

Supplier: Reheis, Inc.

	Ingredient:	Wt.%
A.	Aluminum Chlorohydrate Solution (**Reach 501**)	36.00
B.	Alcohol SD40	30.00
C.	Propylene Glycol	5.00
D.	Deionized Water	28.80
E.	PEG-40 Hydrogenated Castor Oil	0.20
F.	Fragrance	q.s.
		100.00

Mixing Procedure:

Combine Part B and Part D and mix using overhear stirring. Add Part C and mix 10 minutes. Add Part A slowly to the batch. Continue stirring until uniform. Pre-mix Part F into Part E. Disperse this mixture into the batch while using rapid mixing. Mix 5 minutes.

ENHANCED EFFICACY SUSPENSION ROLL-ON

Supplier: Reheis, Inc.

	Ingredient:	Wt.%
A	Aluminum Zirconium Tetrachlorohydrex GLY (**Reach AZP-701**)	20.00
B	Cyclomethicone(and)Quaternium-18 Hectorite (and)Propylene Carbonate	13.50
C	Cyclomethicone	66.00
D	Silicon Dioxide	0.50
E	Fragrance	q.s.
		100.00

Mixing Procedure:
Mix Part B and Part C with an overhead mixer for 20 minutes. Add Part A and mix for 15 minutes. Add Parts D and E and mix for 10 minutes. Homogenize for 3 minutes and pour into suitable containers.

DRY ANTIPERSPIRANT STICK

Supplier: RHEOX, Inc.

	Ingredient:	Wt.%
A	Cyclomethicone	39.70
B	Stearyl Alcohol	22.00
C	Glyceryl Stearate(and)PEG-100 Stearate	2.00
	Titanium Dioxide	0.20
D	Cyclomethicone(and)Quaternium-18 Hectorite (and)Propylene Carbonate (**Bentone Gel VS-5**)	10.00
E	Aluminum Chlorohydrate, Micronized	25.00
	Talc	1.00
F	Fragrance	0.10
		100.00

Mixing Procedure:
Heat Part A to 65°C. With stirring, slowly add Part B, maintaining the temperature at 65°C. When Part B has melted, add the Part C ingredients and mix for 15 minutes. Add Part D and mix for 30 minutes. Add Part E ingredients and mix for 30 minutes. Allow the batch to cool to 55°C, add Part F, mix 5 minutes and pour into suitable containers.

DRY ROLL-ON ANTIPERSPIRANT

Supplier: RHEOX, Inc.

	Ingredient:	Wt.%
	Isopropyl Myristate(and) Stearalkonium Hectorite(and)Propylene Carbonate	
A	(**Bentone Gel IPM**)	15.00
	Alcohol SD40	3.00
	Isopropyl Myristate	2.00
	Cyclomethicone	54.00
B	Aluminum Chlorohydrate, Powder	25.00
C	Fragrance	1.00
		100.00

Mixing Procedure:

Combine the first three Part A ingredients and one-half of the cyclomethicone in a mixing kettle. Mix until uniform. Add the remaining Cyclomethicone and continue mixing. Add Part B with agitation until the powder is uniformly dispersed. Add Part C, mix and fill units.

QUICK-DRYING, ROLL-ON ANTIPERSPIRANT

Supplier: RHEOX, Inc.

	Ingredient:	Wt.%
A	Cyclomethicone(and)Quaternium-18 Hectorite (and)Propylene Carbonate (**Bentone Gel VS-5**)	15.00
B	Cyclomethicone	54.00
C	Alcohol SD40A	3.00
D	Isopropyl Myristate	2.00
E	Aluminum Chlorohydrate Powder	25.00
F	Fragrance	1.00
		100.00

Mixing Procedure:

Combine the **Bentone Gel VS-5** with Parts B and C using vigorous agitation. Mix until uniform. Add Part D , then Part E with constant agitation. Mix until E is uniformly distributed. Add Part F.

TALC SPRAY ANTIPERSPIRANT

Supplier: RHEOX, Inc.

	Ingredient:	Wt.%
A	Isopropyl Myristate(and)Stearalkonium Hectorite(and)Propylene Carbonate (**Bentone Gel IPM**)	8.00
	Alcohol SDA40	2.00
	Isopropyl Myristate	1.50
	Cyclomethicone	5.00
B	Aluminum Chlorohydrate Powder	6.00
	Talc	2.00
C	Fragrance	0.50
	Propellant A-46	75.00
		100.00

Mixing Procedure:

Mix the Part A ingredients thoroughly using medium shear equipment. Add the Part B ingredients slowly and mix thoroughly. Add Fragrance and mix. Fill, vacuum crimp and charge aerosol cans with A-46.

ROLL-ON ANTIPERSPIRANT

Supplier: Southern Clay Products, Inc.

	Ingredient:	Wt.%
A	Montmorillonite (**Gelwhite GP**)	2.00
	Water	26.00
	Mineral Oil(and)Lanolin Alcohol	5.00
	PPG-15 Stearyl Ether	5.00
B	Isopropyl Myristate	15.00
	Sorbitan Oleate	2.00
	Polysorbate 80	2.00
	Glyceryl Stearate(and)PEG-100 Stearate	3.00
C	Aluminum Chlorohydrate, 50%	40.00
D	Preservatives	q.s.
		100.00

Mixing Procedure:

Slowly add **Gelwhite GP** to water while agitating with maximum available shear until smooth. Heat to 65-70°C. Combine the Part B ingredients, heat to 65-70°C and add to Part A slowly. Mix until smooth. Add Part C to Parts A + B with slow speed mixing until smooth and uniform. Add desired preservative and mix until uniform.

ROLL-ON ANTIPERSPIRANT LOTION

Supplier: R.T. Vanderbilt Company, Inc.

	Ingredient:	Wt.%
	Magnesium Aluminum Silicate (**Veegum HV**)	1.00
A	Water	49.60
	Hydroxypropyl Methylcellulose, Medium Visc.	0.40
	Alcohol SD40	8.00
B	Cyclomethicone	3.00
	PPG-15 Stearyl Ether	1.00
	Oleth-10	1.00
C	Aluminum Chlorohydrate, 50%	36.00
	Preservative	q.s.
		100.00

Mixing Procedure:

Slowly add **Veegum HV** to water while agitating with maximum available shear. Mix until smooth. Add the Hydroxypropyl Methylcellulose slowly and mix until smooth. Avoid incorporating air. Add Parts B and C in order, mixing each addition until smooth and uniform.

WATER-IN-OIL STAY DRY STICK

Supplier: R.T. Vanderbilt Company, Inc.

	Ingredient:	Wt.%
	Magnesium Aluminum Silicate (**Veegum HV**)	1.00
A	Water	20.00
	Aluminum Chlorohydrate, Impalpable	18.00
	Cyclomethicone	26.00
	Isopropyl Myristate	5.00
B	Stearyl Alcohol	24.00
	Sorbitan Trioleate	5.00
	Polysorbate 85	1.00
	Preservative	q.s.
		100.00

Mixing Procedure:

Slowly add **Veegum HV** to water while agitating with maximum available shear. Mix until smooth. Add Aluminum Chlorohydrate and mix until smooth. Heat Part A, with stirring, to 70°C. Mix and heat Part B ingredients to 65°C. Add Part A to Part B. Mix and cool to 50°C, pour into molds and cool.

ROLL-ON ANTIPERSPIRANT

Supplier: R.T. Vanderbilt Company, Inc.

	Ingredient:	Wt.%
A	Magnesium Aluminum Silicate (**Veegum HV**)	1.00
	Deionized Water	29.00
B	Glyceryl Stearate(and)PEG-100 Stearate	8.00
	Aluminum Zirconium Tetrachlorohydrex GLY	33.00
C	Aluminum Chloride, 32° Baume Solution	5.50
	Aluminum Chlorohydrate, 50% solution	16.50
D	Cyclomethicone	7.00
E	Fragrance, Dye, Preservative	q.s.
		100.00

Mixing Procedure:

Add **Veegum HV** to water at 75°C and mix with maximum available shear until smooth, uniform and completely free of undispersed particles. In another container, heat Part B to 75°C. Heat Part C to 55°C in a third container. Add Part B to Part A with slow mixing and cool to 50°C. Add Part C to Parts A + B. Mix at slow speed until the temperature reaches 30°C. Add Part D and homogenize for 5 minutes. Add Part E and mix until uniform.

CHAPTER 8

ANTIPRURITIC PRODUCTS

Antipruritic OTC drugs are used to relieve the itching associated with skin irritations, rashes, abrasions and the like. The primary active ingredients used in these brand name products are Hydrocortisone, Pramoxine Hydrochloride, Zinc Oxide, Calamine and Menthol.

The brand name products are usually in the form of creams, lotions or ointments.

PART I

BRAND NAME FORMULATIONS

ALMAY ANTI-ITCH

Manufacturer: Almay, Inc.

Active Ingredients & Concentration:
1% Pramoxine Hydrochloride
+ 0.2% Menthol

Product Form:
Lotion

Other Ingredients:
Water, Petrolatum, Glyceryl Stearate, Glycerin, PEG-40 Stearate, Mineral Oil, Steareth-20, Ceteth-20, Cetyl Alcohol, PEG-150 Stearate, Magnesium Aluminum Silicate, Hydroxypropyl Methylcellulose, Benzyl Alcohol.

ANUSOL HC-1

Manufacturer: Parke-Davis

Active Ingredient & Concentration:
Hydrocortisone Acetate Equal to 1% Hydrocortisone

Product Form:
Cream

Other Ingredients:
Diazolidinyl Urea, Methylparaben, Microcrystalline Wax, Mineral Oil, Propylene Glycol, Propylparaben, Sorbitan Sesquioleate, White Petrolatum.

AVEENO ANTI-ITCH

Manufacturer: Rydelle Laboratories

Active Ingredients & Concentration:
1.0% Pramoxine Hydrochloride + 0.3% Camphor + 3.0% Calamine

Product Form:
Cream

Other Ingredients:
Water, Glycerin, Distearyldimonium Chloride, Petrolatum, Oatmeal Flour, Isopropyl Palmitate, Cetyl Alcohol, Dimethicone, Sodium Chloride.

BACTINE

Manufacturer: Miles Inc.

Active Ingredient & Concentration:
0.5% Hydrocortisone

Product Form:
Cream

Other Ingredients:
BHA, BHT, Butylparaben, Carbomer, Cetyl Alcohol, Colloidal Silicon Dioxide, Corn Oil(and)Glyceryl Oleate(and)PropyleneGlycol(and)BHT(and)BHA(and)Propyl Gallate(and)Citric Acid, DEA-Oleth 3 Phosphate, Diisopropyl Sebecate, Edetate Disodium, Glycerin, Hydroxypropyl Methylcellulose 2906, Lanolin Alcohol, Methylparaben, Mineral Oil (and) Lanolin Alcohol, Propylene Glycol Stearate SE, Propylparaben, Purified Water.

BENADRYL, CREAM, REGULAR & MAXIMUM STRENGTH

Manufacturer: Parke-Davis

Active Ingredients & Concentration:
Regular: 1% Diphenhydramine Hydrochloride
Maximum: 2% Diphenhydramine Hydrochloride + 0.1% Zinc Acetate

Product Form:
Cream

Other Ingredients:
Both strengths contain Cetyl Alcohol, Methylparaben, Polyethylene Glycol Monostearate, Propylene Glycol, Purified Water, Aloe Vera, Propylparaben, Diimidazole Urea.

BENADRYL, SPRAY, REGULAR & MAXIMUM STRENGTH

Manufacturer: Parke-Davis

Active Ingredients & Concentration:
Regular: 1% Diphenhydramine Hydrochloride + 0.1% Zinc Acetate
Maximum: 2% Diphenhydramine Hydrochloride + 0.1% Zinc Acetate

Product Form:
Liquid

Other Ingredients:
Alcohol, Aloe Vera, Glycerin, Povidone, Tromethamine, Purified Water.

CALADRYL

Manufacturer: Parke-Davis

Active Ingredients & Concentration:
Cream, Lotion & Spray: 8% Calamine + 1% Diphenhydramine Hydrochloride
Clear Lotion: 1% Diphenhydramine Hydrochloride + 2% Zinc Oxide

Product Forms:
Cream
Lotion
Spray
Clear Lotion

Other Ingredients:
Cream contains Camphor, Cetyl Alcohol, Fragrance, Methylparaben, Propylene Glycol, Propylparaben, Polysorbate 60, Sorbitan Stearate, Purified Water, Cresin White.
Lotion contains Alcohol (2%), Camphor, Fragrances, Glycerin, Carboxymethylcellulose Sodium, Purified Water.
Spray contains Alcohol (9%), Camphor, FD&C Red No. 40, Fragrance, Isobutane, Quaternium-18 Hectorite, Sorbitan Sesquioleate, Talc.
Clear Lotion contains Alcohol (2%), Camphor, Diazolidinyl Urea, Fragrances, Glycerin, Hydroxypropyl Methylcellulose, Methylparaben, Polysorbate 40, Propylene Glycol, Propylparaben, Water, Chlorophyllin Sodium.

CALDECORT

Manufacturer: Fisons Consumer Health

Active Ingredient & Concentration:
Hydrocortisone Acetate Equal to 1% Hydrocortisone

Product Forms:
Cream
Spray

Other Ingredients:
Cream Contains Isopropyl Myristate, Methylparaben, Polysorbate 60, Propylparaben, Purified Water, Sorbitan Monostearate, Sorbitol, Stearic Acid.
Spray contains Alcohol SD 40-B (62.7%), Isobutane, Isopropyl Myristate.

CALDECORT, LIGHT

Manufacturer: Fisons Consumer Health

Active Ingredient & Concentration:
Hydrocortisone Acetate Equal to 0.5% Hydrocortisone

Product Form:
Cream

Other Ingredients:
Aloe Vera Gel, Isopropyl Myristate, Methylparaben, Polysorbate 60, Propylparaben, Purified Water, Sorbitan Monostearate, Sorbitol, Stearic Acid.

CORTAID, REGULAR AND MAXIMUM STRENGTH

Manufacturer: The Upjohn Company

Active Ingredients & Concentration:
Regular Strength: 0.5% Hydrocortisone
Maximum: Strength: 1% Hydrocortisone

Product Forms:
Cream
Spray
Ointment
Lotion

Other Ingredients:
Regular strength cream contains Aloe Vera, Butylparaben, Cetyl Palmitate, Glyceryl Stearate, Methylparaben, Polyethylene Glycol, Stearamidoethyl Diethylamine, Purified Water.
Maximum strength cream contains Butylparaben, Cetyl Alcohol, Glycerin, Methylparaben, Sodium Lauryl Sulfate, Stearic Acid Stearyl Alcohol, Purified Water, White Petrolatum.
Regular and maximum strength sprays contain Alcohol, Glycerin, Methylparaben, Purified Water.
Regular and maximum strength ointments contain Butylparaben, Cholesterol, Methylparaben, Mineral Oil, Microcrystalline Wax, White Petrolatum. Regular Strength ointment also contains Aloe Vera.
Regular and maximum strength lotions contain Butylparaben, Cetyl Palmitate, Glyceryl Monostearate, Methylparaben, Polysorbate 80, Propylene Glycol, Stearamidoethyl Diethylamine, Purified Water.

CORTIZONE-5 AND CORTIZONE FOR KIDS

Manufacturer: Thompson Medical Company, Inc.

Active Ingredient & Concentration:
All forms: 0.5% Hydrocortisone

Product Forms:
Cream
Ointment
Wipe

Other Ingredients:
Cream contains Aluminum Sulfate, Calcium Acetate, Glycerin, Mineral Oil, Methylparaben, Potato Dextrin, Purified Water, Sodium Lauryl Sulfate, White Petrolatum. May also contain Cetearyl Alcohol, Propylparaben, Sodium C12-15 Alcohols Sulfate, Synthetic Beeswax, White Wax, Aloe vera gel.
Ointment contains White Petrolatum
Wipes contain Methylparaben. Octoxanol 9, Propylene Glycol, Propylparaben, Purified Water.

CORTIZONE-10

Manufacturer: Thompson Medical Company, Inc.

Active Ingredient & Concentration:
Both forms:1.0% Hydrocortisone

Product Forms:
Cream
Ointment

Other Ingredients:
Cream contains Aluminum Sulfate, Calcium Acetate, Cetearyl Alcohol, Glycerin, Mineral Oil, Methylparaben, Potato Dextrin, Water, Sodium Lauryl Sulfate, White Petrolatum, White Wax. May also contain Propylparaben, Aloe Vera Gel, Sodium C12-15 Alcohols Sulfate.
Ointment contains White Petrolatum.

CORTIZONE-10, SCALP ITCH FORMULA

Manufacturer: Thompson Medical Company, Inc.

Active Ingredient & Concentration:
1.0% Hydrocortisone

Product Form:
Liquid

Other Ingredients:
Alcohol SD 40-2 (60% v/v), Benzyl Alcohol, Propylene Glycol, Purified Water.

DERMAREST DRICORT

Manufacturer: Del Pharmaceuticals, Inc.

Active Ingredient & Concentration: Hydrocortisone Acetate Equal to 1% Hydrocortisone

Product Form: Cream

Other Ingredients: Caprylic/Capric Triglyceride, Colloidal Silicon Dioxide, Polyethylene, White Petrolatum.

DOMEBORO

Manufacturer: Miles, Inc.

Active Ingredients & Concentration: 1191mg Aluminum Sulfate + 938mg Calcium Acetate/powder packet

Product Form: Powder

Other Ingredients: Dextrin.

ITCH-X

Manufacturer: B. F. Ascher & Company

Active Ingredients & Concentration: 1% Pramoxine Hydrochloride + 10% Benzyl Alcohol

Product Form: Gel

Other Ingredients: Aloe Vera Gel, Carbomer 934, Diazolidinyl Urea, FD&C Blue No. 1, Methylparaben, Propylene Glycol, Propylparaben, Alcohol SD40, Styrene/Acrylate Copolymer, Triethanolamine, Water.

PREPARATION H

Manufacturer: Whitehall Laboratories, Inc.

Active Ingredient & Concentration: 1% Hydrocortisone

Product Form; Cream

Other Ingredients:
BHA, Cellulose Gum, Cetyl Alcohol, Citric Acid, Edetate Disodium, Glycerin, Glyceryl Oleate, Glyceryl Stearate, Lanolin, Methylparaben, Petrolatum, Propyl Gallate, Propylene Glycol, Propylparaben, Simethicone, Sodium Benzoate, Sodium Lauryl Sulfate, Stearyl Alcohol, Water, Xanthan Gum.

SCALPICIN

Manufacturer: Combe, Inc.

Active Ingredient & Concentration: 1% Hydrocortisone

Product Form: Liquid

Other Ingredients:
Menthol, Propylene Glycol, Alcohol SD40, Water.

PART II

SUPPLIER'S FORMULATIONS

ANTIPRUITIC LOTION

Supplier: Amerchol Corp.

	Ingredient:	Wt.%
A	Hydrocortisone Acetate	0.50
B	Mineral Oil(and)Lanolin Alcohol (**Amerchol L-101**)	3.00
	Cetyl Alcohol(and)Acetylated Lanolin Alcohol (**Acetulan**)	1.00
C	Choleth-24(and)Ceteth-24 (**Solulan C-24**)	0.30
	Stearic Acid XXX	3.00
	Glyceryl Stearate	2.00
	Triethanolamine, 98%	1.00
D	Methyl Gluceth-10 (**Glucam E-10**)	5.00
	Water	84.20
E	Preservative	q.s.
		100.00

Mixing Procedure:

Wet Part A with Part B. Add the Part C ingredients and heat to 75°C. Heat Part D to 75°C. Add Part D to Parts A+B+C. Add Part E to Part D or A+B+C+D as appropriate. Cool to room temperature with moderate agitation. Pass through a homogenizer.

CALAMINE LOTION

Supplier: American Colloid Company

Ingredient:	Wt.%
Calamine	7.83
Zinc Oxide	7.83
Glycerin	1.96
Water	23.08
Calcium Hydroxide Solution	55.80
Purified Bentonite (**Polargel NF**)	3.44
Carboxymethylcellulose	0.06
	100.00

Mixing Procedure:

Prepare a saturated Calcium Hydroxide solution using 3gm of $Ca(OH)_2$ in 1000ml of purified water, mixing vigorously for 1 hr. Decant the clear supernatant liquid for use in the formula. Add the balance of the water. Add the **Polargel NF** and Carboxymethylcellulose to the above solution with rapid mixing for 15 minutes. In a separate vessel, blend the Calamine and Zinc Oxide. Add the Glycerin and mix until uniform. Begin adding the aqueous solution with mixing until it is all blended into the lotion.

HYDROCORTISONE GEL

Supplier: BASF Corp.

	Ingredient:	Wt.%
A	PEG-40 Hydrogenated Castor Oil (**Cremophor RH 40**)	20.00
	Ceteareth-25 (**Cremophor A 25**)	15.00
B	Hydrocortisone Acetate	1.00
C	Water	63.90
	Preservative	0.10
		100.00

Mixing Procedure:

Mix and heat Part A ingredients to 70°C. Add Part B to Part A and disperse thoroughly. Mix and heat Part C ingredients to 70°C. Add Part C to Parts A + B and mix thoroughly. Continue to mix slowly while cooling to room temperature.

CALAMINE LOTION

Supplier: Croda, Inc.

	Ingredient:	Wt.%
A	Calamine Powder	6.50
	PPG-2 Myristyl Ether Propionate (**Crodamol PMP**)	4.00
	Emulsifying Wax NF (**Polawax**)	0.50
	DEA Oleth-3 Phosphate (**Crodafos N3N**)	0.50
	Camphor	1.00
B	Glycerin	4.00
	Carboxymethylcellulose Sodium	2.00
	Deionized Water	77.50
C	Ethanol SDA-40	4.00
		100.00

Mixing Procedure:

Disperse the Calamine Powder into part of the **Crodamol PMP** with mixing. Add the remaining Part A ingredients with mixing and heat to 70-75°C. Disperse the Carboxymethylcellulose Sodium in the Water with mixing and heat to 75-80°C. Mix the Glycerin in this solution. Add Part B to Part A with mixing and cool to 45°C. Add Part C with mixing and adjust the pH to 7.0-8.0 with 10% Citric Acid solution.

CALAMINE CREAM

Supplier: FMC Corp.

	Ingredient:	Wt.%
A	Microcrystalline Cellulose (**Avicel**® **RC-591**)	2.00
B	Glycerin	10.00
C	Methylparaben	0.18
D	Propylparaben	0.02
E	Glyceryl Stearate(and)PEG-100 Stearate	10.00
F	Cetyl alcohol	2.50
G	Zinc oxide	5.00
H	Calamine	5.00
I	Distilled water	65.30
		100.00

Mixing Procedure:

Mix Part B with the Part I and heat to 75°C. Add Parts C and D. Mix until dissolved. Propeller, paddle or anchor type mixers may be used. While mixing, maintain the temperature at 75°C and gradually add the **Avicel RC-591**. Continue the mixing at 75°C for 15 minutes or until the **Avicel RC-591** is homogeneously dispersed. Remove heat. Mix and slowly add the Parts E and F melted together at 75°C. Mix well. When the temperature drops to 60°C-65°C, gradually add Parts G and H. Mix well until the powders are homogeneously dispersed. Adequate powder dispersion was attained using a propeller mixer. If dispersion is not found to be adequate then, after adjustment to weight and when still warm, the fluid cream may be passed through a homogenizer at low pressure or a colloid mill at wide gap setting. Adjust to theoretical weight with warm, distilled water and continue mixing until the cream congeals.

CALAMINE AND ZINC OXIDE LOTION

Supplier: FMC Corp.

	Ingredients:	Wt.%
A	Methylparaben	0.10
B	Propylparaben	0.01
C	Deionized Water	80.99
D	Microcrystalline Cellulose (**Avicel RC-591NF**)	0.80
E	Carboxymethylcellulose Sodium, Medium Visc.	0.10
F	Glycerin	2.00
G	Calamine Powder	8.00
H	Zinc Oxide	8.00
		100.00

Mixing Procedure:

Mix Parts A and B Part C. Slowly add Part D and mix until dispersed. Add Part E and mix until dissolved. Add Part F and mix. Slowly add the Parts G and H in the order shown, mixing each until homogeneous. Pass the batch through a colloid mill at 0.01 inch clearance for 5 minutes.

ANTI-ALLERGENIC CREAM

Supplier: Gattefossé s.a.

	Ingredient:	Wt.%
	PEG-6 Stearate(and)Glyceryl Stearate(and) PEG-32 Stearate (**Tefose 63**)	18.00
A	Caprylic/Capric Triglycerides (**Labrafac Lipophile Wl 1349**)	5.00
	Mineral Oil	10.00
	Sodium Methylparaben	0.05
	Sorbic Acid	0.05
	Demineralized Water	65.90
B	Hydrocortisone Acetate	1.00
		100.00

Mixing Procedure:

Mix and heat the Part A ingredients to 75°C. Cool slowly while stirring. At 30°C, add Part B.

ANTI-INFLAMMATORY CREAM

Supplier: Gattefossé s.a.

	Ingredient:	Wt.%
A	Glyceryl Stearate(and)PEG-75 Stearate (**Gelot 64**)	15.00
	Stearic Acid	0.50
	Mineral Oil	8.00
	Demineralized Water	66.50
B	Salicylic Acid	10.00
		100.00

Mixing Procedure:

Mix and heat Part A ingredients to 75°C. Allow to cool with gentle stirring. At 30 °C, add Salicylic Acid. Homogenize, if necessary.

BABY CREAM

Supplier: Gattefossé s.a.

	Ingredient:	Wt.%
A	Trilaneth-4 Phosphate(and)Glyceryl Stearate (and)PEG-2 Stearate (**Sedefos 75**)	20.00
	Hydrogenated Palm/Palm Kernel Oil PEG-6 Esters (**Labrafil M 2130 CS**)	3.00
	Sweet Almond Oil	5.00
	Mineral Oil	10.00
	Sorbic Acid	0.07
	Sodium Methylparaben	0.03
	Demineralized Water	51.80
	Zinc Oxide	8.00
B	**Soluble Collagen**	2.00
C	Fragrance	0.10
		100.00

Mixing Procedure:

Mix and heat the Part A ingredients to 75°C. Cool with stirring. At 30°C, add Parts B and C. Homogenize, if necessary.

BABY CREAM, W/O

Supplier: Gattefossé s.a.

	Ingredient:	Wt.%
	Base WL 2586	15.00
	Sweet Almond Oil	6.00
A	Mineral Oil	47.00
	Petrolatum	5.00
	Methylparaben	0.10
	Carbomer 934, 1% Aqueous Solution	10.00
	Demineralized Water	15.00
B	Zinc Oxide	1.00
	Sodium Borate	0.75
	Fragrance	0.15
		100.00

Mixing Procedure:

Mix Part A ingredients and B ingredients in separate containers. Heat each Part to 75°C. Add Part B to Part A with constant stirring. Keep the mix at 75°C for 5 minutes. Cool while constantly stirring.

BURN CREAM

Supplier: Gattefossé s.a.

	Ingredient:	Wt.%
	Glyceryl Stearate SE (**Monthybase**)	12.00
	Octyldodecyl Myristate (**M.O.D.**)	8.00
	Apricot Kernel Oil PEG-6 Esters	
A	(**Labrafil M 1944 CS**)	2.00
	Sodium Methylparaben	0.05
	Sodium Propylparaben	0.05
	Sorbic Acid	0.05
	Demineralized Water	76.75
B	Avocado Oil	1.00
C	Fragrance	0.10
		100.00

Mixing Procedure:

Mix and heat the Part A ingredients to 75°C. Cool slowly with stirring. At 30°C, add Part B and then Part C

WOUND OINTMENT

Supplier: Henkel Corp.

	Ingredient:	Wt.%
	Cetearyl Alcohol(and)PEG-40 Castor Oil (and)Sodium Cetearyl Sulfate (**Emulgade F**)	12.00
	Petrolatum	18.00
A	Oleyl Oleate (**Cetiol**)	6.00
	Mineral Oil, Light	6.00
	Zinc Oxide	10.00
	Talc	10.00
B	Water	37.00
	Propylene Glycol(and)Diazolidinyl Urea (and)Methylparaben(and)Propylparaben	1.00
		100.00

Mixing Procedure:

Mix and heat the Part A ingredients to 70-75°C. Mix and heat the Part B ingredients to 70-75°C. While stirring, add Part B to Part A. Begin cooling. Continue stirring until the batch reaches 30°C and homogenize.

FOOT BALSAM

Supplier: Hüls America, Inc.

	Ingredient:	Wt.%
	Glyceryl Stearate SE (**Imwitor 960**)	7.00
A	Stearic Acid	5.00
	Cetyl Alcohol	1.00
	Caprylic/Capric Triglyceride (**Miglyol 812**)	9.00
	Sorbitol	5.00
B	Preservative	q.s.
	Water	69.60
C	Triethanolamine	0.90
D	Mountain Pine Oil	2.00
	Menthol	0.50
		100.00

Mixing Procedure:

Parts A and B are mixed and heated separately to approximately 70°C. Part C is added to Part B and the mixture is emulsified into Part A. Part D is added at approximately 30°C.

CALAMINE LOTION

Supplier: Southern Clay Products Company

	Ingredient:	Wt.%
	Montmorillonite (**Gelwhite H, NF**)	2.00
A	Carboxymethylcellulose Sodium, Low Visc.	0.20
	Water	79.30
	Glycerin	2.50
B	Calamine	8.00
	Zinc Oxide	8.00
		100.00

Mixing Procedure:

Slowly add the **Gelwhite H, NF** to the water while agitating with maximum available shear. Add the Carboxymethylcellulose Sodium and mix at moderate speed until smooth. Add the Glycerin and mix at medium speed. Slowly add the Part B ingredients, with moderate agitation, until smooth and uniform.

HYDROCORTISONE ACETATE CREAM

Supplier: R.T. Vanderbilt Company, Inc.

	Ingredient:	Wt.%
	Magnesium Aluminum Silicate (**Veegum K**)	1.50
	Water	66.49
A	Citric Acid	0.01
	Polysorbate 80	0.50
	Hydrocortisone Acetate	0.50
	C12-15 Alcohols Benzoate	10.00
B	Hydrogenated Polyisobutene	5.00
	C18-36 Acid	4.00
	Glyceryl Stearate(and)PEG-100 Stearate	12.00
C	Preservative	q.s.
		100.00

Mixing Procedure:

Add the **Veegum K** to the water slowly, agitating with maximum available shear until smooth. Add the remaining Part A ingredients in order, mix and heat to 75-80°C. Mix and heat the Part B ingredients to 75-80°C. Add Part B to Part A and mix until smooth and begin cooling. Add Part C at 45°C and mix until uniform. Package.

O/W DIAPER RASH/PRICKLY HEAT LOTION

Supplier: R.T. Vanderbilt Company, Inc.

	Ingredient:	Wt.%
	Magnesium Aluminum Silicate (**Veegum K**)	0.70
A	Water	64.10
	Unimulse C	0.70
	Propylene Glycol	3.00
	Eucalyptus Oil	3.00
	Lanolin Oil	3.00
B	Dimethicone, 350 cs	5.00
	C12-15 Alcohols Benzoate	5.00
	Polysorbate 80	0.50
	Zinc Oxide	5.00
C	Corn Starch	1.00
	Preservative	q.s.
		100.00

Mixing Procedure:

Add the **Veegum K** to the water slowly, agitating with maximum available shear until smooth. Add the remaining Part A ingredients in order, mixing each until uniform. Mix the Part B ingredients until uniform. Add Part B to Part A and mix until uniform. Add the Part C ingredients in order and mix until smooth and uniform.

POISON IVY LOTION

Supplier: R.T Vanderbilt Company, Inc.

	Ingredient:	Wt.%
	Magnesium Aluminum Silicate (**Veegum**)	1.50
A	Carboxymethylcellulose Sodium, Medium Visc.	0.30
	Water	79.65
B	Zirconium Oxide	4.00
	Propylene Glycol	5.00
	Isopropyl Alcohol	8.00
C	Benzocaine	1.50
	Menthol	0.05
	Preservative	q.s.
		100.00

Mixing Procedure:

Dry blend the **Veegum** and Carboxymethylcellulose Sodium and add them to the water slowly while agitating with maximum available shear until smooth. Add Part B to Part A. Add Part C to Parts A + B with rapid mixing.

ZINC OXIDE STICK

Supplier: R.T. Vanderbilt Company, Inc.

	Ingredient:	Wt.%
	Magnesium Aluminum Silicate (**Veegum**)	1.00
A	Water	30.00
	Glycerin	2.50
	Mineral Oil, Light	15.00
	Acetylated Lanolin Alcohol	6.00
	Petrolatum	7.00
	Paraffin	12.50
B	Stearic Acid	2.50
	Zinc Oxide	20.00
	Calcium Stearoyl Lactylate	3.00
	Oleamide DEA	0.5
	Preservative	q.s.
		100.00

Mixing Procedure:

Add the **Veegum** to the water, agitating until smooth and uniform. Stir in Glycerin and heat to 85-90°C. Mix and heat the Part B ingredients to 85-90°C. Add Part A to Part B, mix until uniform and pour into molds to cool.

CHAPTER 9

ASTHMA TREATMENT PRODUCTS

The most commonly used active ingredients in asthma treatment products are Epinephrine and Epinephrine Bitartrate, Recepinephrine, and Theophylline.

The brand name products are available in the form of aerosol sprays and tablets.

PART I

BRAND NAME FORMULATIONS

ASTHMAHALER

Manufacturer: Menley & James Laboratories, Inc.

Active Ingredient & Concentration:	Product Form:
7 mg Epinephrine Bitartrate/ml	Liquid

Other Ingredients:
Cetyl Pyridinium Chloride, Propellants 11, 12, 114, Sorbitan Trioleate.

ASTHMANEFRIN

Manufacturer: Menley & James Laboratories, Inc.

Active Ingredient & Concentration:	Product Form:
Racepinephrine Hydrochloride Equal to 2.25% Epinephrine Base	Liquid

Other Ingredients:
Benzoic Acid, Chlorobutanol, Glycerin, Hydrochloric Acid, Sodium Bisulfite, Sodium Chloride, Water.

BRONKAID

Manufacturer: Sterling Winthrop, Inc.

Active Ingredient & Concentration:	Product Form:
0.5% Epinephrine USP as Nitrate and Hydrochloride Salts	Liquid

Other Ingredients:
Alcohol (33% w/w), Ascorbic Acid, Dichlorodifluoromethane, Purified Water, Dichlorotetrafluoroethane.

BRONKAID SUSPENSION

Manufacturer: Sterling Winthrop, Inc.

Active Ingredient & Concentration:	Product Form:
7.0mg Epinephrine Bitartrate/ml	Suspension

Other Ingredients:
Cetyl Pyridinium Chloride, Dichlorodifluoromethane, Sorbitan Trioleate, Dichlorotetrafluoroethane, Trichloromonofluoromethane.

BRONKAID TABLETS

Manufacturer: Sterling Winthrop, Inc.

Active Ingredients & Concentration:	Product Form:
24mg Ephedrine Sulfate + 100mg Theophylline + 100mg Guaifenesin/Tablet	Tablet

Other Ingredients:
Magnesium Stearate, Magnesium Trisilicate, Microcrystalline Cellulose, Starch.

BRONKOLIXIR

Manufacturer: Sanofi Winthrop Pharmaceuticals

Active Ingredients & Concentration:	Product Form:
12mg Ephedrine Sulfate + 50mg Guaifenesin + 15mg Theophylline + 4mg Phenobarbital/5ml	Liquid

Other Ingredients:
Alcohol (19% v/v), FD&C Red No. 40, Flavors, Glycerin, Purified Water, Saccharin Sodium, Sodium Chloride, Sodium Citrate, Sucrose.

BRONKOTABS

Manufacturer: Sanofi Winthrop Pharmaceuticals

Active Ingredients & Concentration:	Product Form:
24mg Ephedrine Sulfate +100mg Guaifenesin + 100mg Theophylline + 8 mg Phenobarbital/Tablet	Tablet

Other Ingredients:
Magnesium Stearate, Magnesium Trisilicate, Microcrystalline Cellulose, Starch.

PRIMATENE

Manufacturer: Whitehall Laboratories, Inc.

Ingredients & Concentration:	Product Forms:
Liquid: 5.5mg Epinephrine/ml	Liquid
Suspension: 7.0mg Epinephrine Bitartrate/ml	Suspension

Other Ingredients:
Liquid contains Alcohol(34%), Ascorbic Acid, Fluorocarbons, Water.
Suspension contains Fluorocarbons, Sodium Trioleate.

PRIMATENE

Manufacturer: Whitehall Laboratories, Inc.

Ingredients & Concentration:
130mg Theophylline
+ 24mg Ephedrine Hydrochloride/Tablet

Product Form:
Tablet

Other Ingredients:
Croscarmellose Sodium, D&C Yellow No. 10 Lake, FD&C Yellow No. 6 Lake, Magnesium Stearate, Microcrystalline Cellulose, Silica, Starch, Stearic Acid.

PART II

SUPPLIER'S FORMULATIONS

ANTI-ASTHMATIC SUPPOSITORIES

Supplier: Gattefossé s.a.

Ingredient:	Wt.%
Theophylline Monohydrate	12.60
Aminophylline	3.15
Suppocire AIML	84.25
	100.00

Mixing Procedure:
Melt the **Suppocire AIML** at 50°C. Add the active ingredients to the **Suppocire AIML** at 45°C with a turbine mixer. Pour into molds at 34°C, 2.20ml/unit. Air cool to 25°C. Then refrigerate at 15°C.

ANTI-ASTHMATIC CHILDREN'S SUPPOSITORIES

Supplier: Gattefossé s.a.

Ingredient:	Wt.%
Theophylline Monohydrate	8.97
Suppocire AIML	91.03
	100.00

Mixing Procedure:

Melt the **Suppocire AIML** at 50°C. Add the active ingredients to the **Suppocire AIML** at 40°C with a turbine mixer. Pour into molds at 34°C, 1.20ml/unit. Air cool to 25°C. Then refrigerate at 15°C for 30 minutes.

THEOPHYLLINE TABLET I

Supplier: The BF Goodrich Company

	Ingredient	Wt.%
	Theophylline	32.90
A	Microcrystalline Cellulose	56.00
	Carbomer 934P (**Carbopol 974P**)	10.00
	Fumed Silica	0.40
B	Magnesium Stearate	0.70
		100.00

Mixing Procedure:

Direct compression tablets are prepared by first blending the Part A ingredients. Next add Part B, followed by an additional 5 minutes mixing. Direct compression is next using a hydraulic tablet press. Wet granulation can also be used by adding enough water to Part A to form a wet mass when mixed in a Hobart mixer. The mass is screened, successively through 6 and 8 mesh screens. It is then tray dried at 50°C. Screen, successively through 12 and 16 mesh screens. Just prior to tableting, add Part B and blend for 5 minutes in a V-shaped blender. Form tablets in a tablet press.

THEOPHYLLINE TABLET II

Supplier: The BF Goodrich Company

	Ingredient	Wt.%
	Theophylline	32.90
A	Lactose, Anhydrous	56.00
	Carbomer 934P (**Carbopol 934P**)	10.00
	Fumed Silica	0.40
B	Magnesium Stearate	0.70
		100.00

Mixing Procedure:

Direct compression tablets are prepared by first blending the Part A ingredients. Next add Part B, followed by an additional 5 minutes mixing. Direct compression is next, using a hydraulic tablet press.

CHAPTER 10

COLD AND ALLERGY PRODUCTS

This chapter bears testimony to the fact that medical science has yet to find a cure for the common cold. There are 135 brand name cold and allergy relief products contained herein, the largest OTC drug category in this book.

Their active ingredients fall into five classes; decongestants, expectorants, cough suppressants, antihistimines and analgesics/antipyritics. These are used in various combinations, concentrations and forms to produce a myriad of cold and allergy treatments. Products that are primarily analgesics/antipyretics are included in chapter 16.

The FDA Advisory Review Panel has recommended that the following active ingredients be classified as Category I for use in OTC cold and allergy medications:

Decongestants	Antihistamines
Ephedrine(Various Salts)	Doxylamine Succinate
Phenylephrine HCl	Pyrilamine Maleate
Naphazoline HCl	Diphenhydramine HCl
Oxymetazoline HCl	Phenyltoloxamine Citrate
Xylometazoline HCL	Thonzylamine HCl
Phenylephrine	Pheniramine Maleate
Phenylpropanolamine	Brompheniramine Maleate
& Salts	Chlorpheniramine Maleate
Pseudoephedrine	Phenindamine Tartrate

Cough Suppressants	Expectorant
Codeine	Guaifenesin
Dextromethorphan HBr	
Diphenhydramine HCl	

PART I

BRAND NAME FORMULATIONS

ACTIFED

Manufacturer: Burroughs Wellcome Company

Active Ingredients & Concentration:	Product Forms:
60mg Pseudoephedrine Hydrochloride + 2.5mg Triprolidine Hydrochloride/Tablet or 10ml of syrup	Syrup Tablet

Other Ingredients:

Tablets contain Flavor, Hydroxypropyl Methylcellulose, Lactose, Magnesium Stearate, Polyethylene Glycol, Potato Starch, Povidone, Sucrose, Titanium Dioxide.

Syrup contains Methylparaben, Sodium benzoate, D&C yellow No. 10, Glycerin, Purified Water, Sorbitol.

ACTIFED, PLUS

Manufacturer: Burroughs Wellcome Company

Active Ingredients & Concentration:	Product Forms:
500mg Acetaminophen +30mg Pseudoephedrine Hydrochloride + 1.25mg Triprolidine Hydrochloride/ Tablet or Caplet	Caplet Tablet

Other Ingredients:

Caplets or tablets contain Carnauba Wax, Crospovidone, FD&C Blue No. 1 Lake, D&C Yellow No. 10 Lake, Hydroxypropyl Methylcellulose, Magnesium Stearate, Microcrystalline Cellulose, Polyethylene Glycol, Polysorbate 80, Povidone, Pregelatinized Corn Starch, Stearic Acid, Titanium Dioxide.

ACTIFED, SINUS DAYTIME

Manufacturer: Burroughs Wellcome Company

Active Ingredients & Concentration:	Product Forms:
325mg Acetaminophen + 30mg Pseudoephedrine Hydrochloride/ Tablet or Caplet	Caplet Tablet

Other Ingredients:
Caplets or tablets contain Carnauba Wax, Crospovidone, Hydroxypropyl Methylcellulose, Magnesium Stearate, Microcrystalline Cellulose, Polyethylene Glycol, Povidone, Pregelatinized Corn Starch, Sodium Starch Glycolate, Stearic Acid, Titanium Dioxide.

ACTIFED, SINUS NIGHTTIME

Manufacturer: Burroughs Wellcome Company

Active Ingredients & Concentration:	Product Forms:
500mg Acetaminophen + 25 mg Diphenhydramine Hydrochloride + 30 mg Pseudoephedrine Hydrochloride/ Tablet or Caplet	Caplet Tablet

Other Ingredients:
Caplets or tablets contain Carnauba Wax, Crospovidone, FD&C Blue No. 1 Lake, Hydroxypropyl Methylcellulose, Methyl Cellulose, Magnesium Stearate, Microcrystalline Cellulose, Polyethylene Glycol, Polysorbate 80, Povidone, Pregelatinized Corn Starch, Sodium Starch Glycolate, Stearic Acid, Titanium Dioxide.

ADVIL, COLD AND SINUS

Manufacturer: Whitehall Laboratories, Inc.

Active Ingredients & Concentration:	Product Form:
200 mg Ibuprofen + 30mg Pseudoephedrine Hydrochloride/ Tablet	Tablet

Other Ingredients:
Carnauba or equivalent wax, Croscarmellose Sodium, Iron Oxides, Methylparaben, Microcrystalline Cellulose, Propylparaben, Silicon Dioxide, Sodium Benzoate, Sodium Lauryl Sulfate, Starch, Stearic Acid, Sucrose, Titanium Dioxide.

AFRIN

Manufacturer: Schering-Plough HealthCare Products

Active Ingredient & Concentration:	Product Form:
120mg Pseudoephedrine Sulfate/ Tablet	Tablet

Other Ingredients:
Acacia, Butylparaben, Calcium Sulfate, Carnauba Wax, Corn Starch, FD&C Blue No. 1, Gelatin, Lactose, Magnesium Stearate, Neutral Soap, Oleic Acid, Povidone, Rosin, Sugar, Talc, White Wax, Zein.

ALKA-SELTZER PLUS, COLD MEDICINE

Manufacturer: Miles, Inc.

Active Ingredients & Concentration:	Product Form:
325mg Aspirin + 24.08mg Phenylpropanolamine Bitartrate + 2mg Chlorpheniramine Maleate/Tablet	Tablet

Other Ingredients:
Citric Acid, Flavors, Sodium Bicarbonate.

ALKA-SELTZER PLUS, SINUS ALLERGY MEDICINE

Manufacturer: Miles, Inc.

Active Ingredients & Concentration: Product Form:
500mg Aspirin Tablet
+ 24.08mg Phenylpropanolamine Bitartrate
+ 2mg Chlorpheniramine Maleate/Tablet

Other Ingredients:
Aspartame, Citric Acid, Flavors, Sodium Bicarbonate, Tableting Aids.

ALKA-SELTZER PLUS, COLD AND COUGH MEDICINE

Manufacturer: Miles, Inc.

Active Ingredients & Concentration: Product Form:
500mg Aspirin Tablet
+ 24.08mg Phenylpropanolamine Bitartrate
+ 2mg Chlorpheniramine Maleate
+ 10mg Dextromethorphan Hydrobromide/Tablet

Other Ingredients:
Aspartame, Citric Acid, Flavor, Sodium Bicarbonate, Tableting Aids,

---

ALKA-SELTZER PLUS, NIGHTTIME COLD MEDICINE

Manufacturer: Miles, Inc.

Active Ingredients & Concentration: Product Form:
500mg Aspirin Tablet
+ 20mg Phenylpropanolamine Bitartrate
+ 2mg Brompheniramine Maleate
+ 10mg Dextromethorphan Hydrobromide/Tablet

Other Ingredients:
Aspartame, Citric Acid, Flavors, Sodium Bicarbonate, Tableting Aids.

ALLEREST, MAXIMUM STRENGTH

Manufacturer: Fisons Consumer Health

Active Ingredients & Concentration: 30 mg Pseudoephedrine Hydrochloride + 2mg Chlorpheniramine Maleate/Tablet

Product Form: Tablet

Other Ingredients: FD&C Blue No. 1 Lake, Dibasic Calcium Phosphate, Magnesium Stearate, Microcrystalline Cellulose, Povidone, Pregelatinized Starch, Sodium Starch Glycolate.

ALLEREST, 12 HR.CAPLETS

Manufacturer: Fisons Consumer Health

Active Ingredients & Concentration: 12mg Chlorpheniramine Maleate + 75mg Phenylpropanolamine Hydrochloride/ Caplet

Product Form: Caplet,S.R.

Other Ingredients: Carnauba Wax, Colloidal Silicon Dioxide, Lactose, Methyl Cellulose, Polyethylene Glycol, Povidone, Red No. 30, Stearic Acid, Titanium Dioxide, Yellow No. 6.

ALLEREST, CHILDREN'S

Manufacturer: Fisons Consumer Health

Active Ingredients & Concentration: 1mg Chlorpheniramine Maleate + 9.4mg Phenylpropanolamine Hydrochloride/ Tablet

Product Form: Tablet, Ch.

Other Ingredients: Calcium Stearate, Citric Acid, Flavor, Magnesium Trisilicate, Mannitol, Saccharin Sodium, Sorbitol.

A.R.M

Manufacturer: Menley & James Laboratories, Inc.

Active Ingredients & Concentration:
4mg Chlorpheniramine Maleate +
+25mg Phenylpropanolamine Hydrochloride/
Caplet

Product Form:
Caplet

Other Ingredients:
Carnauba Wax, D&C Yellow No. 10, FD&C Yellow No. 6, Gelatin, Hydroxypropyl Methylcellulose, Lactose, Magnesium Stearate, Polyethylene Glycol, Sodium Starch Glycolate, Starch, trace amounts of other ingredients.

BENADRYL, DECONGESTANT

Manufacturer: Parke-Davis

Active Ingredients & Concentration:
Elixir: 12.5mg Diphenhydramine Hydrochloride + 30mg Pseudoephedrine Hydrochloride/5ml
Tablet or Capsule: 25mg Diphenhydramine Hydrochloride + 60mg Pseudoephedrine Hydrochloride/Tablet or Capsule

Product Forms:
Elixir
Tablet
Capsule

Other Ingredients:
Elixir contains Alcohol (5%), Color, Glucose, Glycerin, Flavors, Menthol, Saccharin Sodium, Sodium Citrate, Sucrose, Purified Water.
Tablets contain Corn Starch, Croscarmellose Sodium, Dibasic Calcium Phosphate Dihydrate, FD&C Blue No. 1 Aluminum Lake, Hydroxypropyl Methylcellulose, Microcrystalline Cellulose, Polyethylene Glycol, Polysorbate 80, Stearic Acid, Titanium Dioxide, Zinc Stearate.
Capsules contain Calcium Stearate, Lactose, Silica Gel, Color, Gelatin, Glyceryl Monostearate, PEG-200, Ricinoleate, Titanium Dioxide.

BENADRYL 25

Manufacturer: Parke-Davis

Active Ingredient & Concentration:	Product Forms:
25mg Diphenhydramine Hydrochloride/ Tablet or Capsule	Tablet Capsule

Other Ingredients:

Tablets contain Corn Starch, Croscarmellose Sodium, Aluminum Lake, Hydroxypropyl Methylcellulose, Microcrystalline Cellulose, Polyethylene Glycol, Polysorbate 80, Stearic Acid, Titanium Dioxide, Zinc Stearate, Dibasic Calcium Phosphate Dihydrate, D&C Red No. 27.

Capsules contain Magnesium Stearate, Lactose, Color, Gelatin, Glyceryl Monooleate, PEG-200, Ricinoleate, Titanium Dioxide.

BENADRYL, COLD TABLETS & ALLERGY/SINUS CAPLETS

Manufacturer: Parke-Davis

Active Ingredients & Concentration:	Product Forms:
500mg Acetaminophen + 12.5mg Diphenhydramine Hydrochloride + 30 mg Pseudoephedrine Hydrochloride/Tablet or Caplet	Tablet Caplet

Other Ingredients:

Tablets contain Carboxymethylcellulose, Croscarmellose Sodium, Hydroxypropylcellulose, Hydroxypropyl Methylcellulose, Magnesium Stearate, Microcrystalline Cellulose, Polyethylene Glycol, Propylene Glycol, Starch, Stearic Acid, Titanium Dioxide, Zinc Stearate.

Caplets contain Candillila Wax, Carboxymethylcellulose, Corn Starch, Croscarmellose Sodium, Colors, Hydroxypropylcellulose, Hydroxypropyl Methylcellulose, Microcrystalline Cellulose, Polyethylene Glycol, Polysorbate 80, Stearic Acid, Titanium Dioxide, Zinc Stearate.

BENADRYL, COLD NIGHTTIME FORMULA

Manufacturer: Parke-Davis

Active Ingredients & Concentration:
1000mg Acetaminophen
+ 50mg Diphenhydramine Hydrochloride
+ 60mg Pseudoephedrine Hydrochloride/30ml

Product Form:
Liquid

Other Ingredients:
Alcohol(10%), Citric Acid, Edetate Disodium, Flavor, Glycerin, Polyethylene Glycol, Potassium Sorbate, Propyl Gallate, Propylene Glycol, Sodium Benzoate, Sodium Citrate, Saccharin Sodium, Purified Water, Colors.

BENYLIN

Manufacturer: Parke-Davis

Active Ingredient & Concentration:
12.5mg Diphenhydramine Hydrochloride/5ml

Product Form:
Liquid

Other Ingredients:
Alcohol (5%), Ammonium Chloride, Caramel, Citric Acid, D&C Red No. 30, FD&C Red No. 40, Flavor, Glucose, Glycerin, Menthol, Purified Water, Sodium Citrate, Saccharin Sodium, Sucrose.

BENZEDREX

Manufacturer: Menley & James Laboratories, Inc.

Active Ingredient & Concentration:
0.5% Phenylephrine Hydrochloride

Product Form:
Liquid

Other Ingredients:
Benzalkonium Chloride, Citric Acid, Methylparaben, Sodium Chloride, Sodium Citrate, Water.

BENZEDREX, 12 HOUR

Manufacturer: Menley & James Laboratories, Inc.

Active Ingredient & Concentration: 0.05% Oxymetazoline Hydrochloride

Product Form: Liquid, S.R.

Other Ingredients:
Benzalkonium Chloride, Glycine, Phenylmercuric Acetate, Sorbitol, Water.

BROMFED

Manufacturer: Muro Pharmaceutical, Inc.

Active Ingredients & Concentration: 2mg Brompheniramine Maleate + 30mg Pseudoephedrine Hydrochloride/5ml

Product Form: Liquid

Other Ingredients:
Citric Acid, FD+C Yellow No. 6, Flavor, Glycerin, Methylparaben, Sodium Benzoate, Sodium Citrate, Saccharin Sodium, Sorbitol, Sucrose, Purified Water.

CEROSE-DM

Manufacturer: Wyeth-Ayerst Laboratories

Active Ingredients & Concentration: 15mg Dextromethorphan Hydrobromide + 4mg Chlorpheniramine Maleate + 10mg Phenylephrine Hydrochloride/5ml

Product Form: Liquid

Other Ingredients:
Alcohol (2.4%), Flavor, Citric Acid, Edetate Disodium, FD&C Yellow No. 6, Saccharin Sodium, Sodium Benzoate, Sodium Citrate, Sodium Propionate, Water.

CHERACOL, NASAL SPRAY

Manufacturer: Roberts Pharmaceutical Corp.

Active Ingredient & Concentration: 0.05% Oxymetazoline Hydrochloride

Product Form: Liquid

Other Ingredients:
Benzalkonium Chloride, Glycine, Sodium Hydroxide, Phenylmercuric Acetate, Sorbitol, Artificial Flavor, Purified Water.

CHERACOL, SINUS

Manufacturer: Roberts Pharmaceutical Corp.

Active Ingredients & Concentration: 6mg Dexbrompheniramine Maleate + 120mg Pseudoephedrine Sulfate/Tablet

Product Form: Tablet

Other Ingredients:
Acacia, Calcium Carbonate, Carnauba Wax, Sugar, Colors, Gelatin, Hydrogenated Castor Oil, Magnesium Stearate, Methylparaben, Povidone, Propylparaben, Shellac, Sodium Benzoate, Sucrose, Talc, Titanium Dioxide.

CHERACOL, THROAT SPRAY

Manufacturer: Roberts Pharmaceutical Corp.

Active Ingredient & Concentration: 1.4% Phenol

Product Form: Liquid

Other Ingredients:
Alcohol(12.5%), Citric Acid, FD&C Red No. 40, Flavor, Glycerin, Propylene Glycol, Sodium Citrate, Saccharin Sodium, Sorbitol, Purified Water.

CHERACOL-D

Manufacturer: Roberts Pharmaceutical Corp.

Active Ingredients & Concentration:	Dosage Form:
10mg Dextromethorphan Hydrobromide + 100mg Guaifenesin/5ml	Liquid

Other Ingredients:
Alcohol (4.75%), Benzoic Acid, FD&C Red No. 40, Flavors, Fragrances, Fructose, Glycerin, Propylene Glycol, Sodium Chloride, Sucrose, Purified Water.

CHERACOL, PLUS

Manufacturer: Roberts Pharmaceutical Corp.

Active Ingredients & Concentration:	Product Form:
25mg Phenylpropanolamine + 20mg Dextromethorphan Hydrobromide + 4mg Chlorpheniramine Maleate/15ml	Liquid

Other Ingredients:
Alcohol(8%), Flavors, Glycerin, Methylparaben, Propylene Glycol, Propylparaben, FD&C Red No. 40, Sodium Chloride, Sorbitol, Purified Water.

CHLOR-TRIMETON, ALLERGY

Manufacturer: Schering-Plough HealthCare Products

Active Ingredients & Concentration:	Product Forms:
Tablets: 4mg Chlorpheniramine Maleate/Tablet	Tablet, S.R.
S. R. Tablets: 8 or 12mg Chlorpheniramine Maleate/Tablet	Liquid
Liquid: 2mg Chlorpheniramine Maleate/5ml	Tablet

Other Ingredients:
Tablets contain Corn Starch, D&C Yellow No. 10 Aluminum Lake, Lactose, Magnesium Stearate.
Sustained release tablets contain Acacia, Butylparaben, Calcium Phosphate, Calcium Sulfate, Carnauba Wax, Corn Starch, Colors, Lactose, Magnesium Stearate, Neutral Soap, Oleic Acid, Potato Starch, Rosin, Sugar, Talc, White Wax, Zein.
Liquid contains Alcohol (5%), Benzaldehyde, FD&C Green No. 3, FD&C Yellow No. 6, Flavor, Glycerin, Menthol, Methylparaben, Propylene Glycol, Propylparaben, Sugar, Vanillin, Water.

CHLOR-TRIMETON, ANTIHISTAMINE & DECONGESTANT

Manufacturer: Schering-Plough HealthCare Products

Active Ingredients & Concentration:	Product Forms:
Tablets: 4mg Chlorpheniramine Maleate + 60mg Pseudoephedrine Sulfate/Tablet	Tablet
S. R, tablet: 8mg Chlorpheniramine Maleate + 120mg Pseudoephedrine Sulfate/Tablet	Tablet, S.R.

Other Ingredients:

Tablets contain Corn Starch, FD&C Blue No. 1, Lactose, Magnesium Stearate, Povidone.

Sustained release tablets contain Acacia, Butylparaben, Calcium Sulfate, Carnauba Wax, Corn Starch, Colors, Gelatin, Lactose, Magnesium Stearate, Neutral Soap, Oleic Acid, Povidone, Rosin, Sugar, Talc, White Wax, Zein.

CHLOR-TRIMETON, NON-DROWSY

Manufacturer: Schering-Plough HealthCare Products

Active Ingredient & Concentration:	Product Form:
60mg Pseudoephedrine Sulfate/Tablet	Tablet

Other Ingredients:

Corn Starch, Colors, Lactose, Magnesium Stearate, Povidone.

COMTREX

Manufacturer: Bristol-Myers Products

Active Ingredients & Concentration:	Dosage Forms:
Tablet & Caplet: 325mg Acetaminophen + 30mg Pseudoephedrine Hydrochloride + 2mg Chlorpheniramine Maleate + 10mg Dextromethorphan Hydrobromide/ Tablet or Caplet	Tablet Caplet Gelcap
Gelcaps: 325mg Acetaminophen +12.5mg Phenylpropanolamine Hydrochloride + 2mg Chlorpheniramine Maleate + 10mg Dextromethorphan Hydrobromide/Gelcap	

Other Ingredients:

Tablets contain Corn Starch, D&C Yellow No. 10 Lake, FD&C Red No. 40 Lake, Magnesium Stearate, Methylparaben, Propylparaben, Stearic Acid. May also contain Povidone.

Caplets contain Benzoic Acid, Carnauba Wax, Corn Starch, D&C Yellow No. 10 Lake, FD&C Red No. 40 Lake, Hydroxypropyl Methylcellulose, Magnesium Stearate, Methylparaben, Mineral Oil, Polysorbate 20, Povidone, Propylene Glycol, Propylparaben, Simethicone Emulsion, Sorbitan Monolaurate, Stearic Acid, Titanium Dioxide.

Gelcaps contain D&C Yellow No. 10, FD&C Red No. 40, Gelatin, Glycerin, Propylene Glycol, Povidone, Polyethylene Glycol, Silicon Dioxide, Sorbitol, Titanium Dioxide, Water.

COMTREX, COUGH FORMULA

Manufacturer: Bristol-Myers Products

Active Ingredients & Concentration:	Product Form:
200mg Guaifenesin + 60mg Pseudoephedrine Hydrochloride + 20mg Dextromethorphan Hydrobromide + 500mg Acetaminophen/20ml	Liquid

Other Ingredients:

Alcohol (20%v/v), Citric Acid, FD&C Red No. 40, Flavor, Menthol, Povidone, Saccharin Sodium, Sodium Citrate, Sucrose, Water.

COMTREX, ALLERGY-SINUS

Manufacturer: Bristol-Myers Products

Active Ingredients & Concentration:
500mg Acetaminophen
+ 30mg Pseudoephedrine Hydrochloride
+ 2mg Chlorpheniramine Maleate/
Tablet or Caplet

Product Forms:
Tablet
Caplet

Other Ingredients:
Benzoic Acid, Carnauba Wax, Corn Starch, Colors, Hydroxypropyl Methylcellulose, Mineral Oil, Polysorbate 20, Povidone, Propylene Glycol, Simethicone Emulsion, Sodium Citrate, Sorbitan Monolaurate, Stearic Acid, Titanium Dioxide. May also contain Crospovidone, Erythorbic Acid, Magnesium Stearate, Methylparaben, Microcrystalline Cellulose, Polysorbate 80, Propylparaben, Silicon Dioxide, Wood Cellulose.

COMTREX, DAYTIME FORMULA

Manufacturer: Bristol-Myers Products

Active Ingredients & Concentration:
325mg Acetaminophen
+ 30mg Pseudoephedrine Hydrochloride
+ 10mg Dextromethorphan Hydrobromide/Caplet

Product Form:
Caplet

Other Ingredients:
Corn Starch, D&C Yellow No. 10 Lake, FD&C Red No. 40 Lake, Magnesium Stearate, Methylparaben, Propylparaben, Stearic Acid, Benzoic Acid, Carnauba Wax, Hydroxypropyl Methylcellulose, Mineral Oil, Polysorbate 20, Povidone, Propylene Glycol, Simethicone Emulsion, Sorbitan Monolaurate, Titanium Dioxide.

COMTREX, NIGHTTIME FORMULA

Manufacturer: Bristol-Myers Products

Active Ingredients & Concentration:	Product Form:
325mg Acetaminophen + 30mg Pseudoephedrine Hydrochloride + 10mg Dextromethorphan Hydrobromide + 2mg Chlorpheniramine Maleate/Tablet	Tablet

Other Ingredients:
Benzoic Acid, Carnauba Wax, Corn Starch, Hydroxypropyl Methylcellulose, Magnesium Stearate, Methylparaben, Povidone, Propylparaben, Simethicone Emulsion, Stearic Acid, Titanium Dioxide, Color, Flavor, Polyethylene Glycol, Polysorbate 80, Saccharin Sodium.

COMTREX, NON-DROWSY FORMULA

Manufacturer: Bristol-Myers Products

Active Ingredients & Concentration:	Product Form:
325mg Acetaminophen + 30mg Pseudoephedrine Hydrochloride + 10mg Dextromethorphan Hydrobromide/Caplet	Caplet

Other Ingredients:
Corn Starch, D&C Yellow No. 10 Lake, FD&C Red No. 40 Lake, Magnesium Stearate, Methylparaben, Propylparaben, Stearic Acid, Benzoic Acid, Carnauba Wax, Hydroxypropyl Methylcellulose, Mineral Oil, Polysorbate 20, Povidone, Propylene Glycol, Simethicone Emulsion, Sorbitan Monolaurate, Titanium Dioxide.

CONGESPRIN

Manufacturer: Bristol-Myers Products

Active Ingredients & Concentration:	Product Form:
81mg Acetaminophen + 1.25mg Phenylephrine Hydrochloride/Tablet	Tablet

Other Ingredients:
Calcium Stearate, D&C Red No. 30 Aluminum Lake, D&C Yellow No. 10 Aluminum Lake, Ethyl Cellulose, Flavor, Mannitol, Microcrystalline Cellulose, Polyethylene, Saccharin Calcium, Sucrose.

CONGESTAC

Manufacturer: Menley & James Laboratories, Inc.

Active Ingredients & Concentration: 400mg Guaifenesin + 60mg Pseudoephedrine Hydrochloride/Caplet

Product Form: Caplet

Other Ingredients:
Croscarmellose Sodium, Hydroxypropyl Methylcellulose, Magnesium Stearate, Microcrystalline Cellulose, Polyethylene Glycol, Povidone, Silicon Dioxide, Starch and trace amounts of other ingredients.

CONTAC, REGULAR & MAXIMUM STRENGTH DECONGESTANT

Manufacturer: SmithKline Beecham Consumer Brands

Active Ingredients & Concentration:
Regular Strength Capsule:
75mg Phenylpropanolamine Hydrochloride + 8mg Chlorpheniramine Maleate/Capsule
Maximum Strength Caplets:
75mg Phenylpropanolamine Hydrochloride + 12mg Chlorpheniramine Maleate/Capsule

Product Forms: Capsule, S.R.; Caplet, S.R.

Other Ingredients:
Regular Strength Capsules contain Benzyl Alcohol, Butylparaben, Carboxymethylcellulose Sodium, Colors, Edetate Calcium Disodium, Gelatin, Methylparaben, Pharmaceutical Glaze, Polysorbate 80, Propylparaben, Sodium Lauryl Sulfate, Sodium Propionate, Starch, Sucrose and other ingredients.
Maximum Strength Caplets contain Acetylated Monoglycerides, Carnauba Wax, Silicon Dioxide, Ethylcellulose, Hydroxypropyl Methylcellulose, Lactose, Stearic Acid, Titanium Dioxide.

CONTAC, DAY & NIGHT COLD & FLU

Manufacturer: SmithKline Beecham Consumer Brands

Active Ingredients & Concentration:	Product Form:
Day Caplets: 650mg Acetaminophen + 60mg Pseudoephedrine Hydrochloride + 30mg Dextromethorphan Hydrobromide/Caplet Night Caplets: 650mg Acetaminophen + 60mg Pseudoephedrine Hydrochloride + 50mg Diphenhydramine Hydrochloride/Caplet	Caplet

Other Ingredients:
Both products contain Carnauba Wax, Colors, Hydroxypropyl Methylcellulose, Magnesium Stearate, Microcrystalline Cellulose, Polyethylene Glycol, Polysorbate 80, Silicon Dioxide, Starch, Stearic Acid, Titanium Dioxide, White Wax.

CONTAC, SEVERE COLD & FLU

Manufacturer: SmithKline Beecham, Inc.

Active Ingredients & Concentration:	Product Forms:
Caplets: 500mg Acetaminophen + 15mg Dextromethorphan Hydrobromide + 12.5mg Phenylpropanolamine Hydrochloride + 2mg Chlorpheniramine Maleate/Caplet Liquid: 500mg Acetaminophen + 15mg Dextromethorphan Hydrobromide + 30mg Pseudoephedrine Hydrochloride + 2mg Chlorpheniramine Maleate/15ml	Caplet Liquid

Other Ingredients:
Caplets contain Cellulose, FD&C Blue No. 1, Hydroxypropyl Methylcellulose, Polyethylene Glycol, Polysorbate 80, Povidone, Sodium Starch Glycolate, Starch, Stearic Acid, Titanium Dioxide.
Liquid contains Alcohol (18.5% by volume), Colors, Dibasic Sodium Phosphate, Flavors, Glycerin, Hydrogenated Glucose, Phosphoric Acid, Potassium Sorbate, Povidone, Saccharin Sodium, Sorbitol, Water.

CORICIDIN

Manufacturer: Schering-Plough HealthCare Products

Active Ingredients & Concentration:
2mg Chlorpheniramine Maleate
+ 325mg Acetaminophen/Tablet

Product Form:
Tablet

Other Ingredients:
Acacia, Butylparaben, Calcium Sulfate, Carnauba Wax, Cellulose, Corn Starch, FD&C Red No. 40 Aluminum Lake, FD&C Yellow No. 6 Aluminum Lake, Lactose, Magnesium Stearate, Povidone, Sugar, Talc, Titanium Dioxide, White Wax.

CORICIDIN 'D'

Manufacturer: Schering-Plough HealthCare Products

Active Ingredients & Concentration:
2mg Chlorpheniramine Maleate
+ 325mg Acetaminophen
+ 12.5mg Phenylpropanolamine Hydrochloride/Tablet

Products Form:
Tablet

Other Ingredients:
Acacia, Butylparaben, Calcium Sulfate, Carnauba Wax, Cellulose, Corn Starch, Magnesium Stearate, Povidone, Sugar, Talc, Titanium Dioxide, White Wax.

DELSYM

Manufacturer: Fisons Consumer Health

Active Ingredient & Concentration:
30mg Dextromethorphan Hydrobromide/ 5ml

Product Form:
Liquid

Other Ingredients:
Citric Acid, Ethylcellulose, FD&C Yellow No. 6, Flavor, High Fructose Corn Syrup, Methylparaben, Polyethylene Glycol 3350, Polysorbate 80, Propylene Glycol, Propylparaben, Purified Water, Sucrose, Tragacanth, Vegetable Oil, Xanthan Gum.

DIMETAPP

Manufacturer: A.H. Robins Company, Inc.

Active Ingredients & Concentration:	Dosage Forms:
Chewable Tablet: 1mg Brompheniramine Maleate + 6.25mg Phenylpropanolamine Hydrochloride/Tablet	Tablet, Ch.
Elixir: 2mg Brompheniramine Maleate + 12.5mg Phenylpropanolamine Hydrochloride/5ml	Elixir
S. R. Tablets: 12mg Brompheniramine Maleate + 75mg Phenylpropanolamine Hydrochloride/ Sustained Release Tablet	Tablet, S.R.
Tablets & Liqui-Gel: 4mg Brompheniramine Maleate + 25mg Phenylpropanolamine Hydrochloride/Tablet or Gelcap	Liqui-Gel Tablet

Other Ingredients:

Chewable tablets contain Aspertame, Citric Acid, Crospovidone, Colors, Flavor, Glycine, Magnesium Stearate, Mannitol, Microcrystalline Cellulose, Pregelatinized Starch, Silicon Dioxide, Sorbitol, Stearic Acid.

Elixir contains Glycerin, Citric Acid, FD&C Blue No. 1, FD&C Red No. 40, Flavor, Saccharin Sodium, Sodium Benzoate, Sorbitol, Water,

Sustained release tablets contain Acacia, Acetylated Monoglycerides, Calcium Sulfate, Carnauba Wax, Castor Wax or Oil, Citric Acid, Edible Ink, Colors, Gelatin, Magnesium Stearate, Magnesium Trisilicate, Pharmaceutical Glaze, Polysorbates, Povidone, Silicon Dioxide, Stearyl Alcohol, Sucrose, Titanium Dioxide, Wheat Flour, White Wax.

Tablets contain Corn Starch, FD&C Blue No. 1, Magnesium Stearate, Microcrystalline Cellulose,

Liqui-Gels contain Gelatin, Glycerin, Mannitol, Pharmaceutical Glaze, Polyethylene Glycol, Povidone, Propylene Glycol, Sorbitol, Titanium Dioxide, Water, Colors.

DIMETAPP DM

Manufacturer: A.H. Robins Company, Inc.

Active Ingredients & Concentration:
2mg Brompheniramine Maleate
+ 12.5mg Phenylpropanolamine Hydrochloride
+ 10mg Dextromethorphan Hydrobromide/5ml

Product Form: Liquid

Other Ingredients:
Alcohol (2.3%), Citric Acid, FD&C Blue No. 1, FD&C Red No. 40, Flavors, Glycerin, Propylene Glycol, Saccharin Sodium, Sodium Benzoate, Sorbitol, Water.

DOROCOL, CHILDREN'S COUGH, COLD & DECONGESTANT FORMULA

Manufacturer: Sandoz Pharmaceuticals/Consumer Div.

Active Ingredients & Concentration:
Cough Syrup: 15mg Pseudoephedrine Hydrochloride + 50mg Guaifenesin + 5mg Dextromethorphan Hydrobromide/5ml
Decongestant: 15mg Pseudoephedrine Hydrochloride/5ml
Cold Formula: 15mg Pseudoephedrine Hydrochloride + 1mg Chlorpheniramine Maleate/5ml

Product Form: Liquids

Other Ingredients:
Cough syrup contains Benzoic Acid, FD&C Blue No. 1, Edetate Disodium, Flavors, Glycerin, Propylene Glycol, Purified Water, FD&C Red No. 40, Sodium Hydroxide, Sucrose, Tartaric Acid.
Decongestant contains Benzoic Acid, Edetate Disodium, Flavors, Purified Water, Sodium Hydroxide, Sorbitol, Sucrose, FD&C Yellow No. 6, D&C Yellow No. 10.
Cold Formula contains Benzoic Acid, FD&C Blue No.1, Flavors, Purified Water, Colors, Sorbitol, Sucrose. May also contain Sodium Hydroxide.

DRISTAN, DECONGESTANT & MENTHOL NASAL SPRAYS

Manufacturer: Whitehall Laboratories, Inc.

Active Ingredients & Concentration:
Decongestant and Menthol sprays:
0.5% Phenylephrine Hydrochloride
+ 0.2% Chlorpheniramine Maleate

Dosage Form:
Liquids

Other Ingredients:
Decongestant spray contains Alcohol (0.4%), Benzalkonium Chloride, Eucalyptol, Hydroxypropyl Methylcellulose, Menthol, Sodium Chloride, Sodium Phosphate, Preservative, Water.
Menthol spray contains Benzalkonium Chloride Camphor, Polysorbate 80, Eucalyptol, Hydroxypropyl Methylcellulose, Menthol, Methyl Salicylate, Sodium Chloride, Sodium Phosphate, Preservative, Water.

DRISTAN, COLD

Manufacturer: Whitehall Laboratories, Inc.

Active Ingredients & Concentration:
5mg Phenylephrine Hydrochloride
+ 2mg Chlorpheniramine Maleate
+ 325 mg Acetaminophen/Tablet

Product Form:
Tablet

Other Ingredients:
Calcium Stearate, Croscarmellose Sodium, Colors, Hydroxypropyl Methylcellulose, Microcrystalline Cellulose, Polyethylene Glycol, Povidone, Starch, Stearic Acid.

DRISTAN, COLD, MAXIMUM STRENGTH, NO DROWSINESS

Manufacturer: Whitehall Laboratories, Inc.

Active Ingredients & Concentration:
30mg Pseudoephedrine Hydrochloride
+ 500mg Acetaminophen/Tablet

Product Form:
Caplet

Other Ingredients:
Calcium Stearate, Croscarmellose Sodium, Colors, Hydrogenated Vegetable Oil, Hydroxypropyl Methylcellulose, Stearic Acid, Microcrystalline Cellulose, Pharmaceutical Glaze, Polyethylene Glycol, Povidone, Starch, Titanium Dioxide.

DRISTAN, COLD, MAXIMUM STRENGTH

Manufacturer: Whitehall Laboratories, Inc.

Active Ingredients & Concentration:
30mg Pseudoephedrine Hydrochloride
+ 2mg Brompheniramine Maleate
+ 500mg Acetaminophen/Gelcap

Product Form:
Gelcap

Other Ingredients:
Calcium Stearate, Croscarmellose Sodium, Colors, EDTA, Gelatin, Glycerin, Hydrogenated Vegetable Oil, Hydroxypropyl Methylcellulose, Iron Oxide, Lecithin, Stearic Acid Microcrystalline Cellulose, Pharmaceutical Glaze, Polyethylene Glycol, Povidone, Simethicone, Starch, Titanium Dioxide, Triacetin.

DRISTAN, COLD, JUICE MIX-IN

Manufacturer: Whitehall Laboratories, Inc.

Active Ingredients & Concentration:
60mg Pseudoephedrine Hydrochloride
+ 20mg Dextromethorphan Hydrobromide
+ 500mg Acetaminophen/Packet of Powder

Product Form:
Powder

Other Ingredients:
Hydroxypropyl Methylcellulose, Povidone, Silica, Starch, Stearic Acid, Sucrose, Tricalcium Phosphate.

DRISTAN, ALLERGY

Manufacturer: Whitehall Laboratories, Inc.

Active Ingredients & Concentration:
60mg Pseudoephedrine Hydrochloride
+ 4mg Brompheniramine Maleate/Caplet

Dosage Form:
Caplet

Other Ingredients:
Ammonium Hydroxide, Calcium Stearate, Colors, Hydrogenated Vegetable Oil, Hydroxypropyl Methylcellulose, Iron Oxide, Microcrystalline Cellulose, Pharmaceutical Glaze, Polyethylene Glycol, Polysorbate 80, Potassium Hydroxide, Propylene Glycol, Silica, Titanium Dioxide.

DRIXORAL

Manufacturer: Schering-Plough HealthCare Products

Active Ingredients & Concentration:
2mg Brompheniramine Maleate
+ 30 mg Pseudoephedrine Sulfate/5ml

Product Form:
Liquid

Other Ingredients:
Citric Acid, D&C Red No. 33, FD&C Yellow No. 6, Flavor, Propylene Glycol, Sodium Benzoate, Sodium Citrate, Sorbitol, Sugar, Water.

DRIXORAL, COLD AND ALLERGY

Manufacturer: Schering-Plough HealthCare Products

Active Ingredients & Concentration:
120mg Pseudoephedrine Sulfate
+ 6mg Dexbrompheniramine Maleate

Product Form:
Tablet, S.R

Other Ingredients:
Acacia, Butylparaben, Calcium Sulfate, Carnauba Wax, Corn Starch, Colors, Gelatin, Lactose, Magnesium Stearate, Neutral Soap, Oleic Acid, Povidone, Rosin, Sugar, Talc, White Wax, Zein.

DRIXORAL, NON-DROWSY

Manufacturer: Schering-Plough HealthCare Products

Active Ingredient & Concentration:
120mg Pseudoephedrine Sulfate/Tablet

Product Form:
Tablet, S.R.

Other Ingredients:
Acacia, Butylparaben, Calcium Sulfate, Carnauba Wax, Corn Starch, FD&C Blue No. 1, Gelatin, Lactose, Magnesium Stearate, Neutral Soap, Oleic Acid, Povidone, Rosin, Sugar, Talc, White Wax, Zein.

DRIXORAL, COLD AND FLU

Manufacturer: Schering-Plough HealthCare Products

Active Ingredients & Concentration:
60mg Pseudoephedrine Sulfate
+ 3mg Dexbrompheniramine Maleate
+ 500mg Acetaminophen/Tablet

Product Form:
Tablet, S.R.

Other Ingredients:
Calcium Phosphate, Carnauba Wax, Colors, Hydroxypropyl Methylcellulose, Magnesium Stearate, Methylparaben, Polyethylene Glycol, Propylparaben, Stearic Acid.

4-WAY, COLD

Manufacturer: Bristol-Myers Products

Active Ingredients & Concentration:
325mg Acetaminophen
+ 12.5mg Phenylpropanolamine Hydrochloride
+ 2mg Chlorpheniramine Maleate/Tablet

Product Form:
Tablet

Other Ingredients:
Corn Starch, Corn Starch Pregelatinized, Microcrystalline Cellulose, Sodium Starch Glycolate, Stearic Acid, Sucrose.

4-WAY, FAST ACTING SPRAY

Manufacturer: Bristol-Myers Products

Active Ingredients & Concentration:
0.5% Phenylephrine Hydrochloride
+ 0.05% Naphazoline Hydrochloride
+ 0.2% Pyrilamine Maleate

Product Form:
Liquid

Other Ingredients:
Benzalkonium Chloride, Boric Acid, Sodium Borate, Water. May also contain Menthol, Poloxamer 188, Polysorbate 80, Camphor, Eucalyptol.

4-WAY, LONG LASTING SPRAY

Manufacturer: Bristol-Myers Products

Active Ingredient & Concentration: 0.05% Oxymetazoline Hydrochloride

Product Form: Liquid, S.R.

Other Ingredients:
Phenylmercuric Acetate (0.002%), Water, Benzalkonium Chloride, Glycine, Sorbitol.

GUAIFED/GUAITAB

Manufacturer: Muro Pharmaceutical, Inc.

Active Ingredients & Concentration:
Liquid: 30mg Pseudoephedrine Hydrochloride + 200mg Guaifenesin/5 ml
Tablet: 60mg Pseudoephedrine Hydrochloride + 400 mg Guaifenesin/Tablet

Product Forms:
Liquid
Tablet

Other Ingredients:
Liquid contains Benzoic Acid, Flavor, Citric Acid, FD&C Red No. 40, Glycerin, Menthol, Polyethylene Glycol, Povidone, Purified Water, Saccharin Sodium, Sodium Citrate, Sorbitol, Vanillin.
Tablets contain Colloidal Silicon Dioxide, Lactose, Magnesium Stearate, Microcrystalline Cellulose, Pharmaceutical Glaze, Sodium Starch Glycolate, Starch, Talc, Colors.

HOLD DM

Manufacturer: Menley & James Laboratories Inc.

Active Ingredient & Concentration: 5mg Dextromethorphan Hydrobromide/ Lozenge

Product Form: Lozenge

Other Ingredients:
Corn Syrup, Colors, Flavor, Vegetable Oil, Sucrose and trace amounts of other ingredients.

ISOCLOR

Manufacturer: Fisons Consumer Health

Active Ingredients & Concentration:	Product Form:
8 mg Chlorpheniramine Maleate	Capsule
+ 120mg Pseudoephedrine Hydrochloride/Capsule	

Other Ingredients:
Castor Oil, Ethylcellulose, Gelatin, Mineral Oil, Silicone Oil, Sugar, White Petrolatum.

MEDI-FLU

Manufacturer: Parke-Davis

Active Ingredients & Concentration:	Product Forms:
Caplets: 500mg Acetaminophen	Caplet
+ 30mg Pseudoephedrine Hydrochloride	Liquid
+ 2mg Chlorpheniramine Maleate	
+ 15mg Dextromethorphan Hydrobromide/Caplet	
Liquid: 60mg Pseudoephedrine Hydrochloride	
+ 4mg Chlorpheniramine Maleate	
+ 30mg Dextromethorphan Hydrobromide	
+ 1000mg Acetaminophen/30 ml	

Other Ingredients:
Caplets contain Cellulose Derivatives, Croscarmellose Sodium, Magnesium Stearate, Carnauba Wax, Polyethylene Glycol, Starch, Stearic Acid.
Liquid contains Alcohol (19%), Citric Acid, FD&C Red No. 40, D&C Red No. 33, Flavors, Glycerin, Propylene Glycol, Sodium Citrate, Sodium Chloride, Saccharin Sodium, Sorbitol, Sugar, Water.

MEDI-FLU, WITHOUT DROWSINESS

Manufacturer: Parke-Davis

Active Ingredients & Concentration:	Product Form:
500 mg Acetaminophen	Caplet
+ 30mg Pseudoephedrine Hydrochloride	
+ 15mg Dextromethorphan Hydrobromide/Caplet	

Other Ingredients:
Candillila Wax, Corn Starch, Croscarmellose Sodium, Colors, Hydroxypropyl Methylcellulose, Microcrystalline Cellulose, Polyethylene Glycol, Polysorbate 80, Titanium Dioxide, Zinc Stearate.

N'ICE

Manufacturer: SmithKline Beecham Consumer Brands

Active Ingredient & Concentration:	Product Form:
5mg Menthol/Lozenge	Lozenge

Other Ingredients:
Sorbitol, Colors, Flavors, may also contain Citric Acid or Tartaric Acid, Saccharin Sodium or Hydrogenated Glucose, Sodium Citrate.

NOVAHISTINE

Manufacturer: SmithKline Beecham Consumer Brands

Active Ingredients & Concentration:	Product Form:
2mg Chlorpheniramine Maleate + 5mg Phenylephrine Hydrochloride/5ml	Liquid

Other Ingredients:
Alcohol (5%), D&C Yellow No. 10, FD&C Blue No. 1, Flavors, Glycerin, Sodium Chloride, Sorbitol, Water.

NOVAHISTINE DMX

Manufacturer: SmithKline Beecham Consumer Brands

Active Ingredients & Concentration:	Product Form:
10mg Dextromethorphan Hydrobromide + 100mg Guaifenesin + 30mg Pseudoephedrine Hydrochloride/5ml	Liquid

Other Ingredients:
Alcohol(10%), FD&C Red No. 40, FD&C Yellow No. 6, Flavors, Glycerin, Hydrochloric Acid, Invert Sugar, Saccharin Sodium, Sodium Chloride, Sorbitol, Water.

ORNEX

Manufacturer: Menley & James Laboratories, Inc.

Active Ingredients & Concentration:
30mg Pseudoephedrine Hydrochloride
+ 325mg Acetaminophen/Caplet

Product Form:
Caplet

Other Ingredients:
Cellulose, Crospovidone, FD&C Blue No.1, Hydroxypropyl Methylcellulose, Magnesium Stearate, Microcrystalline Cellulose, Polyethylene Glycol, Polysorbate 80, Povidone, Starch, Titanium Dioxide and trace amounts of other ingredients.

ORNEX, MAXIMUM STRENGTH

Manufacturer: Menley & James Laboratories, Inc.

Active Ingredients & Concentration:
30mg Pseudoephedrine Hydrochloride
+ 500mg Acetaminophen/Caplet

Dosage Form:
Caplet

Other Ingredients:
Cellulose, Crospovidone, Hydroxypropyl Methylcellulose, Magnesium Stearate, Polyethylene Glycol, Polysorbate 80, Povidone, Starch, Titanium Dioxide and trace amounts of other ingredients.

ORNEX, SEVERE COLD FORMULA

Manufacturer: Menley & James Laboratories, Inc.

Active Ingredients & Concentration:
30 mg Pseudoephedrine Hydrochloride
+ 500mg Acetaminophen
+ 15mg Dextromethorphan Hydrobromide/Caplet

Product Form:
Caplet

Other Ingredients:
Cellulose, Crospovidone, Colors, Hydroxypropyl Methylcellulose, Magnesium Stearate, Polyethylene Glycol, Povidone, Starch, Titanium Dioxide and trace amounts of other ingredients.

ORTHOXICOL

Manufacturer: Roberts Pharmaceutical Corp.

Active Ingredients & Concentration:
25mg Phenylpropanolamine
+ 20mg Dextromethorphan Hydrobromide
+ 4mg Chlorpheniramine/15ml

Product Form:
Liquid

Other Ingredients:
Alcohol (8%), Flavors, Glycerin, Methylparaben, Propylene Glycol, Propylparaben, FD&C Red No. 40, Sodium Chloride, Sorbitol, Purified Water.

PEDIACARE, COLD-ALLERGY

Manufacturer: McNeil Consumer Products Company

Active Ingredients & Concentration:
1mg Chlorpheniramine Maleate
+ 15mg Pseudoephedrine Hydrochloride/Tablet

Product Form:
Tablet, Ch.

Other Ingredients:
3mg Phenylalanine/tablet, Aspartame, Cellulose, Citric Acid, Corn Starch, Flavors, Mannitol, Colloidal Silicon Dioxide, Stearic Acid, Color.

PEDIACARE, COUGH-COLD

Manufacturer: McNeil Consumer Products Company

Active Ingredients & Concentration:
Tablet: 1mg Chlorpheniramine Maleate
+ 15mg Pseudoephedrine Hydrochloride
+ 5mg Dextromethorphan Hydrobromide/Tablet
Liquid: 1mg Chlorpheniramine Maleate
+ 15 mg Pseudoephedrine Hydrochloride
+ 5mg Dextromethorphan Hydrobromide/5ml

Product Forms:
Tablet, Ch.
Liquid

Other Ingredients:
Chewable Tablets contain 3 mg Phenylalanine/tablet, Aspartame, Cellulose, Citric Acid, Starch, Flavors, Magnesium Stearate, Magnesium Trisilicate, Mannitol, Color.
Liquid contains Benzoic Acid, Citric Acid, Flavors, Glycerin, Polyethylene Glycol, Propylene Glycol, Sodium Benzoate, Sorbitol, Sucrose, Purified Water, Colors.

PEDIACARE, NIGHTREST

Manufacturer: McNeil Consumer Products Company

Active Ingredients & Concentration:
1mg Chlorpheniramine Maleate
+ 15mg Pseudoephedrine Hydrochloride
+ 7.5mg Dextromethorphan Hydrobromide/5ml

Product Form:
Liquid

Other Ingredients:
Benzoic Acid, Citric Acid, Flavors, Glycerin, Polyethylene Glycol, Propylene Glycol, Sodium Benzoate, Sorbitol, Sucrose, Purified Water, Colors.

PEDIACARE, INFANT DROPS

Manufacturer: McNeil Consumer Products Company

Active Ingredients & Concentration:
7.5mg Pseudoephedrine Hydrochloride/0.8ml

Product Form:
Liquid

Other Ingredients:
Benzoic Acid, Citric Acid, Flavors, Glycerin, Polyethylene Glycol, Propylene Glycol, Sodium Benzoate, Sorbitol, Sucrose, Purified Water, Color.

PYRROXATE

Manufacturer: Roberts Pharmaceutical Corp.

Active Ingredients & Concentration:
25mg Phenylpropanolamine Hydrochloride
+ 500mg Acetaminophen
+ 4mg Chlorpheniramine Hydrochloride/Capsule

Product Form:
Capsule

Other Ingredients:
Benzyl Alcohol, Butylparaben, Colors, Erythrosine Sodium, Gelatin, Glycerin, Magnesium Stearate, Methylparaben, Propylparaben, Sodium Lauryl Sulfate, Sodium Propionate, Starch, Talc.

ROBITUSSIN-CF

Manufacturer: A.H. Robins Company, Inc.

Active Ingredients & Concentration:
100mg Guaifenesin
+ 12.5mg Phenylpropanolamine Hydrochloride
+ 10 mg Dextromethorphan Hydrobromide/5ml

Product Form:
Liquid

Other Ingredients:
Alcohol(4.75%), Citric Acid, FD&C Red No. 40, Flavors, Glycerin, Propylene Glycol, Saccharin Sodium, Sodium Benzoate, Sorbitol, Water.

ROBITUSSIN, COUGH CALMERS

Manufacturer: A.H. Robins Company, Inc.

Active Ingredient & Concentration:
5 mg Dextromethorphan Hydrobromide/ Lozenge

Product Form:
Lozenge

Other Ingredients:
Corn Syrup, FD&C Red No. 40., Flavors, Glycerin, Gum Acacia, Sucrose.

ROBITUSSIN-DM

Manufacturer: A.H. Robins Company, Inc.

Active Ingredients & Concentration:
100mg Guaifenesin
+ 10mg Dextromethorphan Hydrobromide/5ml

Product Form:
Liquid

Other Ingredients:
Citric Acid, FD&C Red No. 40, Flavors, Glucose, Glycerin, High Fructose Corn Syrup, Saccharin Sodium, Sodium Benzoate, Water.

ROBITUSSIN, MAXIMUM STRENGTH COUGH FORMULA

Manufacturer: A.H. Robins Company, Inc.

Active Ingredient & Concentration:
15mg Dextromethorphan Hydrobromide /5ml

Product Form:
Liquid

Other Ingredients:
Citric Acid, FD&C Red No. 40, Flavors, Glucose, Glycerin, High Fructose Corn Syrup, Saccharin Sodium, Sodium Benzoate, Water.

ROBITUSSIN, MAXIMUM STRENGTH COUGH & COLD

Manufacturer: A.H. Robins Company, Inc.

Active Ingredients & Concentration:
15mg Dextromethorphan Hydrobromide
+ 30 mg Pseudoephedrine Hydrochloride/5ml

Product Form:
Liquid

Other Ingredients:
Alcohol(1.4%), Citric Acid, FD&C Red No. 40, Flavors, Glucose, Glycerin, High Fructose Corn Syrup, Saccharin Sodium, Sodium Benzoate, Water.

ROBITUSSIN, NIGHT RELIEF COUGH, COLD, FLU

Manufacturer: A.H. Robins Company, Inc.

Active Ingredients & Concentration:
30mg Dextromethorphan Hydrobromide
+ 60mg Pseudoephedrine Hydrochloride
+ 50mg Pyrilamine Maleate
+ 650mg Acetaminophen/m

Product Form:
Liquid

Other Ingredients:
Citric Acid, FD&C Red No. 40, FD&C Blue No. 1, Flavors, Glycerin, Propylene Glycol, Saccharin Sodium, Sodium Benzoate, Sorbitol, Water and other ingredients.

ROBITUSSIN-PE

Manufacturer: A.H. Robins Company, Inc.

Active Ingredients & Concentration: 100mg Guaifenesin + 30mg Pseudoephedrine Hydrochloride/5ml

Product Form: Liquid

Other Ingredients: Alcohol(1.4%), Citric Acid, FD&C Red No. 40, Flavors, Glucose, Glycerin, High Fructose Corn Syrup, Saccharin Sodium, Sodium Benzoate, Water.

ROBITUSSIN, PEDIATRIC COUGH & COLD

Manufacturer: A.H. Robins Company, Inc.

Active Ingredients & Concentration: 7.5mg Dextromethorphan Hydrobromide + 15mg Pseudoephedrine Hydrochloride/5 ml

Product Form: Liquid

Other Ingredients: Citric Acid, FD&C Red No. 40, Flavors, Glycerin, Propylene Glycol, Saccharin Sodium, Sodium Benzoate, Sorbitol, Water.

ROBITUSSIN, PEDIATRIC COUGH SUPPRESSANT

Manufacturer: A.H. Robins Company, Inc.

Active Ingredient & Concentration: 7.5mg Dextromethorphan Hydrobromide/ 5ml

Product Form: Liquid

Other Ingredients: Citric Acid, FD&C Red No. 40, , Flavors, Glycerin, Propylene Glycol, Saccharin Sodium, Sodium Benzoate, Sorbitol, Water.

RYNA

Manufacturer: Wallace Laboratories

Active Ingredients & Concentration: 2 mg Chlorpheniramine Maleate + 30mg Pseudoephedrine Hydrochloride/5ml

Product Form: Liquid

Other Ingredients: Flavor, Glycerin, Malic Acid, Sodium Benzoate, Sorbitol, Purified Water.

RYNA-C

Manufacturer: Wallace Laboratories

Active Ingredient & Concentration:
10mg Codeine Phosphate/5ml

Product Form:
Liquid

Other Ingredients:
Flavor, Glycerin, Malic Acid, Sodium Benzoate, Saccharin Sodium, Sorbitol, Purified Water.

RYNA-CX

Manufacturer: Wallace Laboratories

Active Ingredients & Concentration:
100mg Guaifenesin
+ 30mg Pseudoephedrine Hydrochloride
+ 10mg Codeine Phosphate/5ml

Product Form:
Liquid

Other Ingredients:
Flavor, Glycerin, Glycine, Malic Acid, Povidone, Propylene Glycol, Sorbitol, Saccharin Sodium, Purified Water.

ST. JOSEPH, COLD TABLETS FOR CHILDREN

Manufacturer: Schering-Plough HealthCare Products

Active Ingredients & Concentration:
80mg Acetaminophen
+ 3.125mg Phenylpropanolamine Hydrochloride/Tablet

Product Form:
Tablet

Other Ingredients:
Cellulose, FD&C Yellow No. 6 Aluminum Lake, Flavor, Mannitol, Silica, Saccharin Sodium, Zinc Stearate.

ST. JOSEPH, COUGH SUPPRESSANT FOR CHILDREN

Manufacturer: Schering-Plough HealthCare Products

Active Ingredients & Concentration:
7.5 mg Dextromethorphan Hydrobromide/5ml

Product Form:
Liquid

Other Ingredients:
Caramel, Citric Acid, Flavor, Glycerin, Methylparaben, Propylparaben, Sodium Benzoate, Sodium Citrate, Sucrose, Water.

SINAREST

Manufacturer: Fisons Consumer Health

Active Ingredients & Concentration:
325mg Acetaminophen
+ 2mg Chlorpheniramine Maleate
+ 30mg Pseudoephedrine Hydrochloride/Tablet

Product Form:
Tablet

Other Ingredients:
Magnesium Stearate, Microcrystalline Cellulose, Povidone, Pregelatinized Starch, Colors.

SINAREST, EXTRA STRENGTH

Manufacturer: Fisons Consumer Health

Active Ingredients & Concentration:
500mg Acetaminophen
+ 2mg Chlorpheniramine Maleate
+ 30mg Pseudoephedrine Hydrochloride/Tablet

Product Form:
Tablet

Other Ingredients:
Magnesium Stearate, Microcrystalline Cellulose, Povidone, Pregelatinized Starch, Sodium Starch Glycolate, Colors
.

SINAREST, NO DROWSINESS

Manufacturer: Fisons Consumer Health

Active Ingredients & Concentration:
500mg Acetaminophen
+ 30mg Pseudoephedrine Hydrochloride/Tablet

Product Form:
Tablet

Other Ingredients:
Magnesium Stearate, Microcrystalline Cellulose, Povidone, Pregelatinized Starch, Sodium Starch Glycolate.

SINGLET

Manufacturer: SmithKline Beecham Consumer Brands

Active Ingredients & Concentration:
60mg Pseudoephedrine Hydrochloride
+ 4mg Chlorpheniramine Maleate
+ 650mg Acetaminophen/Tablet

Product Form:
Tablet

Other Ingredients:
Colors, Hydroxypropylcellulose, Hydroxypropyl Methylcellulose 2910, Magnesium Stearate, Microcrystalline Cellulose, Polyethylene Glycol 8000, Pregelatinized Corn Starch, Sodium Starch Glycolate, Sucrose, Titanium Dioxide.

SINUTAB, SINUS ALLERGY

Manufacturer: Parke-Davis

Active Ingredients & Concentration:
30 mg Pseudoephedrine Hydrochloride
+ 2 mg Chlorpheniramine Maleate
+ 500 mg Acetaminophen/Tablet or Caplet

Product Forms:
Tablet
Caplet

Other Ingredients:
Tablets contain Carboxymethyl Starch, Microcrystalline Cellulose, Corn Starch, Croscarmellose Sodium, Hydroxypropylcellulose, Stearic Acid, Zinc Stearate, Colors.
Caplets contain Microcrystalline Cellulose, Corn Starch, Croscarmellose Sodium, Magnesium Stearate, Hydroxypropylcellulose, Stearic Acid, Hydroxypropyl Methylcellulose, Polyethylene Glycol, Simethicone, Sodium Starch Glycolate, Titanium Dioxide, Zinc Stearate, Colors.

SUDAFED, COUGH SYRUP

Manufacturer: Burroughs Wellcome Company

Active Ingredients & Concentration:
15 mg Pseudoephedrine Hydrochloride
+ 5mg Dextromethorphan Hydrobromide
+ 100mg Guaifenesin/5ml

Product Form:
Liquid

Other Ingredients:
Alcohol(2.4%), Methylparaben, Sodium Benzoate, Citric Acid, Colors, Flavor, Glycerin, Purified Water, Saccharin Sodium, Sodium Chloride, Sucrose.

SUDAFED

Manufacturer: Burroughs Wellcome Company

Active Ingredient & Concentration:	Product Form:
30 or 60mg Pseudoephedrine Hydrochloride/Tablet	Tablet

Other Ingredients:

Both strengths contain Acacia, Carnauba Wax, Dibasic Calcium Phosphate, Magnesium Stearate, Pharmaceutical Glaze, Polysorbate 60, Stearic Acid, Talc, Titanium Dioxide. 30 mg tablets also contain Colors, Potato Starch, Povidone, Sodium Benzoate. 60 mg tablets also contain Hydroxypropyl Methylcellulose, Sodium Starch Glycolate, Sucrose.

SUDAFED, PLUS

Manufacturer: Burroughs Wellcome Company

Active Ingredients & Concentration:	Product Forms:
Liquid: 30mg Pseudoephedrine Hydrochloride + 2mg Chlorpheniramine Maleate/5ml	Liquid
Tablet: 60mg Pseudoephedrine Hydrochloride + 4mg Chlorpheniramine Maleate/Tablet	Tablet

Other Ingredients:

Liquid contains Methylparaben, Sodium Benzoate, Citric Acid, Colors, Flavor, Glycerin, Purified Water, Sucrose.

Tablets contain Lactose, Magnesium Stearate, Potato Starch, Povidone.

SUDAFED, SEVERE COLD FORMULA

Manufacturer: Burroughs Wellcome Company

Active Ingredients & Concentration:	Product Forms:
500mg Acetaminophen	Tablet
+ 15mg Dextromethorphan Hydrobromide	Caplet
+ 30mg Pseudoephedrine Hydrochloride/ Tablet or Caplet	

Other Ingredients:
Both forms contain contain Carnauba Wax, Crospovidone, Hydroxypropyl Methylcellulose, Magnesium Stearate, Microcrystalline Cellulose, Polyethylene Glycol, Povidone, Corn Starch, Stearic Acid, Titanium Dioxide.

SUDAFED, CHILDREN'S LIQUID

Manufacturer: Burroughs Wellcome Company

Active Ingredient & Concentration:	Product Form:
30mg Pseudoephedrine Hydrochloride/ 5ml	Liquid

Other Ingredients:
Methylparaben, Sodium Benzoate, Citric Acid, FD&C Red No. 40, Flavor, Glycerin, Purified Water, Sorbitol, Sucrose.

TAVIST-1

Manufacturer: Sandoz Pharmaceuticals/Consumer Div.

Active Ingredient & Concentration:	Product Form:
1.34mg Clemastine Fumarate/Tablet	Tablet

Other Ingredients:
Lactose, Povidone, Starch, Stearic Acid, Talc.

TAVIST-D

Manufacturer: Sandoz Pharmaceuticals/Consumer Div.

Active Ingredients & Concentration:	Product Form:
1.34mg Clemastine Fumarate + 75mg Phenylpropanolamine Hydrochloride/ Tablet	Tablet

Other Ingredients:
Colloidal Silicon Dioxide, Dibasic Calcium Phosphate, Lactose, Magnesium Stearate, Methylcellulose, Polyethylene Glycol, Povidone, Starch, Synthetic Polymers, Titanium Dioxide, Color.

TELDRIN

Manufacturer: SmithKline Beecham Consumer Brands

Active Ingredient & Concentration:	Product Form:
12mg Chlorpheniramine Maleate/Capsule	Capsule,S.R.

Other Ingredients
Benzyl Alcohol, Cetylpyridinium Chloride, Colors, Gelatin, Hydrogenated Castor Oil, Silicon Dioxide, Sodium Lauryl Sulfate, Starch, Sucrose and trace amounts of other ingredients.

THERAFLU, FLU & COLD MEDICINE

Manufacturer: Sandoz Pharmaceuticals/Consumer Div.

Active Ingredients & Concentration:	Product Form:
650mg Acetaminophen + 60mg Pseudoephedrine Hydrochloride + 4mg Chlorpheniramine Maleate + 20mg Dextromethorphan Hydrobromide/Powder Packet	Powder

Other Ingredients:
Ascorbic Acid, Citric Acid, Flavor, Sodium Citrate, Sucrose, Titanium Dioxide, Tribasic Calcium Phosphate, Pregelatinized Starch, Colors.

THERAFLU, MAXIMUM STRENGTH, NIGHTTIME

Manufacturer: Sandoz Pharmaceuticals/Consumer Div.

Active Ingredients & Concentration:
1000mg Acetaminophen
+ 60mg Pseudoephedrine Hydrochloride
+ 4mg Chlorpheniramine Maleate
+ 30mg Dextromethorphan Hydrobromide/Powder Packet

Product Form:
Powder

Other Ingredients:
Ascorbic Acid, Citric Acid, Flavor, Maltol, Sodium Citrate, Sucrose, Titanium Dioxide, Tribasic Calcium Phosphate, Pregelatinized Starch, Colors Silicon Dioxide.

TRIAMINIC, ALLERGY, COLD & CHEWABLES

Manufacturer: Sandoz Pharmaceuticals/Consumer Div.

Active Ingredients & Concentration:
Allergy Tablet: 4mg Chlorpheniramine Maleate + 25mg Phenylpropanolamine Hydrochloride/Tablet
Cold Tablet: 2mg Chlorpheniramine Maleate + 12.5mg Phenylpropanolamine Hydrochloride/Tablet
Chewables: 0.5mg Chlorpheniramine Maleate + 6.25 mg Phenylpropanolamine Hydrochloride/Tablet

Product Forms:
Tablet, Ch. Tablet

Other Ingredients:
Allergy tablets contain Calcium Stearate, Calcium Sulfate, Colloidal Silicon Dioxide, Methylcellulose, Methylparaben, Microcrystalline Cellulose, Polyethylene Glycol, Povidone, Pregelatinized Starch, Titanium Dioxide, Color.
Cold tablets contain Calcium Stearate, Flavor, Lactose, Colloidal Silicon dioxide, Methylcellulose, Methylparaben, Microcrystalline Cellulose, Polyethylene Glycol, Povidone, Pregelatinized Starch, Colors, Saccharin Sodium, Titanium Dioxide.
Chewables contain Calcium Stearate, Citric Acid, Flavors, Magnesium Trisilicate, Mannitol, Microcrystalline Cellulose, Saccharin Sodium, Sucrose, Colors.

TRIAMINIC-12

Manufacturer: Sandoz Pharmaceuticals/Consumer Div.

Active Ingredients & Concentration:
75mg Phenylpropanolamine Hydrochloride
+ 12mg Chlorpheniramine Maleate/Tablet

Product Form:
Tablet, S.R.

Other Ingredients:
Carnauba Wax, Colloidal Silicon Dioxide, Lactose, Methylcellulose, Polyethylene Glycol, Povidone, Colors, Stearic Acid, Titanium Dioxide.

TRIAMINIC, EXPECTORANT

Manufacturer: Sandoz Pharmaceuticals/Consumer Div.

Active Ingredients & Concentration:
6.25mg Phenylpropanolamine Hydrochloride
+ 50mg Guaifenesin/5ml

Product Form:
Liquid

Other Ingredients:
Benzoic Acid, Edetate Disodium, Flavors, Purified Water, Saccharin Sodium, Sodium Hydroxide, Sorbitol, Sucrose, Colors.

TRIAMINIC, NITE LIGHT

Manufacturer: Sandoz Pharmaceuticals/Consumer Div.

Active Ingredients & Concentration:
15mg Pseudoephedrine Hydrochloride
+ 1mg Chlorpheniramine Maleate
+ 7.5mg Dextromethorphan Hydrobromide/5ml

Product Form:
Liquid

Other Ingredients:
Benzoic Acid, Citric Acid, Flavors, Propylene Glycol, Purified Water, Sorbitol, Sucrose, Colors, Dibasic Sodium Phosphate.

TRIAMINIC-DM

Manufacturer: Sandoz Pharmaceuticals/Consumer Div.

Active Ingredients & Concentration:	Product Form:
6.25mg Phenylpropanolamine Hydrochloride + 5mg Dextromethorphan Hydrobromide/5ml	Liquid

Other Ingredients:
Benzoic Acid, Flavors, Propylene Glycol, Purified Water, Sodium Chloride, Sorbitol, Sucrose, Colors.

TRIAMINICIN

Manufacturer: Sandoz Pharmaceuticals/Consumer Div.

Active Ingredients & Concentration:	Product Form:
650mg Acetaminophen + 25mg Phenylpropanolamine Hydrochloride + 4 mg Chlorpheniramine Maleate/Tablet	Tablet

Other Ingredients:
Colloidal Silicon Dioxide, Croscarmellose Sodium, Hydroxypropyl Methylcellulose, Lactose, Magnesium Stearate, Methylcellulose, Methylparaben, Polyethylene Glycol, Povidone, Pregelatinized Starch, Colors, Titanium Dioxide.

TRIAMINICOL, MULTISYMPTOM

Manufacturer: Sandoz Pharmaceuticals/Consumer Div.

Active Ingredients & Concentration:	Product Forms:
2mg Chlorpheniramine Maleate + 10mg Dextromethorphan Hydrobromide + 12.5mg Phenylpropanolamine Hydrochloride/Tablet or 10ml	Tablet Liquid

Other Ingredients:
Tablets contain Calcium Stearate, Colloidal Silicon Dioxide, Lactose, Methylcellulose, Methylparaben, Microcrystalline Cellulose, Polyethylene Glycol, Povidone, Pregelatinized Starch, Color, Titanium Dioxide.
Liquid contains Benzoic Acid, Flavors, Propylene Glycol, Purified Water, Color, Saccharin Sodium, Sodium Chloride, Sorbitol, Sucrose.

TYLENOL, MAXIMUM STRENGTH COUGH MEDICATION

Manufacturer: McNeil Consumer Products Company

Active Ingredients & Concentration:	Product Form:
30mg Dextromethorphan Hydrobromide + 1000mg Acetaminophen/20ml	Liquid

Other Ingredients:
Alcohol (10%), Citric Acid, Flavors, Glycerin, Polyethylene Glycol, Purified Water, Sodium Benzoate, Carboxymethylcellulose Sodium, Saccharin Sodium, Sorbitol, Sucrose, Colors.

TYLENOL, MAXIMUM STRENGTH COUGH MEDICATION WITH DECONGESTANT

Manufacturer: McNeil Consumer Products Company

Active Ingredients & Concentration:	Product Form:
30mg Dextromethorphan Hydrobromide + 1000mg Acetaminophen + 60mg Pseudoephedrine Hydrochloride/20ml	Liquid

Other Ingredients:
Alcohol (10%), Citric Acid, Flavors, Glycerin, Polyethylene Glycol, Purified Water, Sodium Benzoate, Carboxymethylcellulose Sodium, Saccharin Sodium, Sorbitol, Sucrose, Colors.

TYLENOL, COLD MEDICATION, NO DROWSINESS

Manufacturer: McNeil Consumer Products Company

Active Ingredients & Concentration:	Product Forms:
Caplets/Gelcaps: 325mg Acetaminophen	Caplet
+ 30mg Pseudoephedrine Hydrochloride	Gelcap
+ 15mg Dextromethorphan Hydrobromide	Powder
Caplet or Gelcap	
Powder: 650mg Acetaminophen	
+ 60mg Pseudoephedrine Hydrochloride	
+ 30mg Dextromethorphan Hydrobromide/Powder	

Other Ingredients:

Caplets contain Cellulose, Glyceryl Triacetate, Hydroxypropyl Methylcellulose, Magnesium Stearate, Sodium Starch Glycolate, Starch, Titanium Dioxide, Colors.

Gelcaps contain Benzyl Alcohol, Butylparaben, Castor Oil, Cellulose, Corn Starch, Edetate, Calcium Disodium, Gelatin, Hydroxypropyl Methylcellulose, Magnesium Stearate, Methylparaben, Propylparaben, Sodium Propionate, Sodium Lauryl Sulfate, Sodium Starch Glycolate, Titanium Dioxide, Colors.

Powder contains Aspartame, Citric Acid, Corn Starch, Flavors, Sodium Citrate, Sucrose, Colors.

TYLENOL, COLD, NIGHT TIME

Manufacturer: McNeil Consumer Products Company

Active Ingredients & Concentration:	Product Form:
50mg Diphenhydramine Hydrochloride	Liquid
+ 60mg Pseudoephedrine Hydrochloride	
+ 650mg Acetaminophen/30ml	

Other Ingredients:

Alcohol (10%), Citric Acid, Flavors, Glycerin, Polyethylene Glycol, Purified Water, Sodium Benzoate, Sucrose, Colors.

TYLENOL, CHILDREN'S COLD

Manufacturer: McNeil Consumer Products Company

Active Ingredients & Concentration:	Product Form:
0.5mg Chlorpheniramine Maleate	Liquid
+ 7.5mg Pseudoephedrine Hydrochloride	Tablet, Ch.
+ 80mg Acetaminophen/Tablet or 2.5ml	

Other Ingredients:

Liquid contains Benzoic Acid, Citric Acid, Flavors, Glycerin, Malic Acid, Polyethylene Glycol, Propylene Glycol, Sodium Benzoate, Sorbitol, Sucrose, Purified Water, Colors.

Chewable Tablets contain Aspartame, Citric Acid, Ethylcellulose, Flavors, Magnesium Stearate, Mannitol, Microcrystalline Cellulose, Pregelatinized Starch, Sucrose, Colors.

TYLENOL, CHILDREN'S, COLD PLUS COUGH

Manufacturer: McNeil Consumer Products Company

Active Ingredients & Concentration:	Product Form:
1mg Chlorpheniramine Maleate	Liquid
+ 15mg Pseudoephedrine Hydrochloride	
+ 160mg Acetaminophen	
+ 5mg Dextromethorphan Hydrobromide/5ml	

Other Ingredients:

Citric Acid, Corn Syrup, Flavors, Polyethylene Glycol, Propylene Glycol, Sodium Benzoate, Carboxymethylcellulose Sodium, Sorbitol, Purified Water, Colors.

TYLENOL, MULTISYMPTOM

Manufacturer: McNeil Consumer Products Company

Active Ingredients & Concentration:	Product Forms:
2mg Chlorpheniramine Maleate + 30mg Pseudoephedrine Hydrochloride + 325mg Acetaminophen + 15 mg + 15mg Dextromethorphan Hydrobromide/Tablet or Caplet	Tablet Caplet

Other Ingredients:

Tablets contain Cellulose, Starch, Magnesium Stearate, Colors.

Caplets contain Cellulose, Glyceryl Triacetate, Hydroxypropyl Methylcellulose, Magnesium Stearate, Sodium Starch Glycolate, Starch, Titanium Dioxide, Colors.

TYLENOL, MAXIMUM STRENGTH, ALLERGY, SINUS

Manufacturer: McNeil Consumer Products Company

Active Ingredients & Concentration:	Product Forms:
500mg Acetaminophen + 2mg Chlorpheniramine Maleate + 30mg Pseudoephedrine Hydrochloride/ Caplet or Gelcap	Caplet Gelcap

Other Ingredients:

Caplets contain Cellulose, Hydroxypropylcellulose, Hydroxypropyl Methylcellulose, Magnesium Stearate, Polyethylene Glycol, Sodium Starch Glycolate, Corn Starch, Titanium Dioxide, Colors.

Gelcaps contain Benzyl Alcohol, Butylparaben, Castor Oil, Cellulose, Edetate Calcium Disodium, Gelatin, Hydroxypropyl Methylcellulose, Magnesium Stearate, Methylparaben, Propylparaben, Sodium Lauryl Sulfate, Sodium Propionate, Sodium Starch Glycolate, Starch, Titanium Dioxide, Colors.

VICKS, EXTRA STRENGTH COUGH DROPS

Manufacturer: Procter & Gamble Company

Active Ingredients & Concentration:	Product Form:
8.4mg Menthol (Menthol Flavor) or 10mg Menthol (Cherry, Peppermint or Honey Lemon Flavor)/Lozenge	Lozenge

Other Ingredients:
Corn Syrup, Colors, Flavors, Sucrose.

VICKS, DAYQUIL

Manufacturer: Procter & Gamble Company

Active Ingredients & Concentration:	Product Form:
Liquid: 650mg Acetaminophen + 20mg Dextromethorphan Hydrobromide + 60mg Pseudoephedrine Hydrochloride + 200mg Guaifenesin/m	Liquid
Gelcaps: 250mg Acetaminophen + 10mg Dextromethorphan Hydrobromide + 30mg Pseudoephedrine Hydrochloride + 100mg Guaifenesin/Gelcap	Gelcap

Other Ingredients:
Liquid contains Citric Acid, FD&C Yellow No. 6, Flavor, Glycerin, Polyethylene Glycol, Propylene Glycol, Purified Water, Saccharin Sodium, Sodium Citrate, Sucrose.
Gelcaps contain Colors, Gelatin, Gelatin Glycerin, Polyethylene Glycol, Povidone, Propylene Glycol, Purified Water.

VICKS FORMULA 44, REGULAR & PEDIATRIC STRENGTH

Manufacturer: Procter & Gamble Company

Active Ingredients & Concentration: Product Form: Liquid

Regular Strength:
30mg Dextromethorphan Hydrobromide/10ml
Pediatric Strength: 15mg Dextromethorphan Hydrobromide/15ml

Other Ingredients:

Regular strength contains Alcohol (10%), Caramel, Carboxymethylcellulose Sodium, Citric Acid, FD&C Red No. 40, Flavor, Sugar, Propylene Glycol, Purified Water, Sodium Citrate.

Pediatric strength contains Carboxymethylcellulose Sodium, Cellulose, Citric Acid, FD&C Red No. 40, Flavors, Glycerin, Polysorbate 80, Potassium Sorbate, Polyethylene Glycol, Purified Water, Sodium Citrate, Sucrose, Sorbitol.

VICKS FORMULA 44D, REGULAR & PEDIATRIC STRENGTH

Manufacturer: Procter & Gamble Company

Active Ingredients & Concentration: Product Form: Liquid

Regular Strength:
30mg Dextromethorphan Hydrobromide + 60mg Pseudoephedrine Hydrochloride/15ml
Pediatric Strength: 15mg Dextromethorphan Hydrobromide + 30mg Pseudoephedrine Hydrochloride/15ml

Other Ingredients:

Regular strength contains Alcohol (10%), Citric Acid, FD&C Red No. 40, Flavor, Glycerin, Propylene Glycol, Purified Water, Saccharin Sodium, Sodium Citrate, Sucrose.

Pediatric strength contains Carboxymethylcellulose Sodium, Cellulose, Citric Acid, FD&C Red No. 40, Flavor, Glycerin, Polysorbate 80, Potassium Sorbate, Polyethylene Glycol, Purified Water, Sodium Citrate, Sorbitol, Sucrose.

VICKS FORMULA 44E, REGULAR & PEDIATRIC STRENGTH

Manufacturer: Procter & Gamble Company

Active Ingredients & Concentration:
Regular Strength:
20 mg Dextromethorphan Hydrobromide
+ 200mg Guaifenesin/15ml
Pediatric Strength: 10 mg Dextromethorphan
Hydrobromide + 100mg Guaifenesin/15ml

Product Form:
Liquid

Other Ingredients:
Regular strength contains Alcohol (10%), Citric Acid, Colors, Flavor, Glycerin, Propylene Glycol, Purified Water, Saccharin Sodium, Sodium Citrate, Sucrose.
Pediatric strength contains Carboxymethylcellulose Sodium, Cellulose, Citric Acid, FD&C Red No. 40, Flavor, Glycerin, Polysorbate 80, Potassium Sorbate, Polyethylene Glycol, Purified Water, Sodium Citrate, Sorbitol, Sucrose.

VICKS FORMULA 44M, REGULAR & PEDIATRIC STRENGTH

Manufacturer: Procter & Gamble Company

Active Ingredients & Concentration:
Regular Strength: 500mg Acetaminophen
+ 30mg Dextromethorphan Hydrobromide
+ 60mg Pseudoephedrine Hydrochloride
+ 4mg Chlorpheniramine Maleate/20ml
Pediatric Strength: 15mg Dextromethorphan
Hydrobromide + 30mg Pseudoephedrine Hydrochloride
+2 mg Chlorpheniramine Maleate/15ml

Product Form:
Liquid

Other Ingredients:
Regular strength contains Alcohol (20%), Citric Acid, Colors, Flavor, Glycerin, Purified Water, Saccharin Sodium, Sodium Benzoate, Sodium Citrate, Sucrose.
Pediatric strength contains Carboxymethylcellulose Sodium, Cellulose, Citric Acid, FD&C Red No. 40, Flavors, Glycerin, Polysorbate 80, Potassium Sorbate, Polyethylene Glycol, Purified Water, Sodium Citrate, Sorbitol, Sucrose.

VICKS NYQUIL

Manufacturer: Procter & Gamble Company

Active Ingredients & Concentration:
Liquid: 1000mg Acetaminophen
+ 30mg Dextromethorphan Hydrobromide
+ 60mg Pseudoephedrine Hydrochloride
+ 12.5mg Doxylamine Succinate/30ml
Liquicap: 250mg Acetaminophen
+ 10mg Dextromethorphan Hydrobromide
+ 30mg Pseudoephedrine Hydrochloride
+ 6.25mg Doxylamine Succinate/Liquicap

Product Forms:
Liquid
Liquicap

Other Ingredients:
Liquid contains Alcohol (10%), Citric Acid, Colors, Flavor, Glycerin, Purified Water, Sodium Citrate, Sucrose. Original flavor also contains Saccharin Sodium, Propylene Glycol, Polyethylene Glycol. Liquicaps contain Colors, Gelatin, Glycerin, Polyethylene Glycol, Povidone, Propylene Glycol, Purified Water. May also contain Edible Ink.

VICKS CHILDREN'S NYQUIL, COUGH & COLD MEDICINE

Manufacturer: Procter & Gamble Company

Active Ingredients & Concentration:
15mg Dextromethorphan Hydrobromide
+ 30mg Pseudoephedrine Hydrochloride
+ 2mg Chlorpheniramine Maleate/15ml

Product Form:
Liquid

Other Ingredients:
Citric Acid, FD&C Red No. 40, Flavor, Potassium Sorbate, Propylene Glycol, Purified Water, Sodium Citrate, Sucrose.

VICKS CHILDREN'S NYQUIL, ALLERGY & COLD MEDICINE

Manufacturer: Procter & Gamble Company

Active Ingredients & Concentration:
+ 30mg Pseudoephedrine Hydrochloride
+ 2mg Chlorpheniramine Maleate/15ml

Product Forms:
Liquid

Other Ingredients:
Citric Acid, Colors, Flavors, Methylparaben, Potassium Sorbate, Propylene Glycol, Purified Water, Sodium Citrate, Sorbitol, Sucrose.

VICKS SINEX, REGULAR STRENGTH

Manufacturer: Procter & Gamble Company

Active Ingredient & Concentration:
0.5% Phenylephrine Hydrochloride

Product Form:
Liquid

Other Ingredients:
Camphor, Eucalyptol, Menthol, Cetylpyridinium Chloride, Potassium Phosphate, Purified Water, Sodium Chloride, Sodium Phosphate, Tyloxapol.

VICKS SINEX, LONG-LASTING

Manufacturer: Procter & Gamble Company

Active Ingredient & Concentration:
0.05% Oxymetazoline Hydrochloride

Product Form:
Liquid, S.R.

Other Ingredients:
Camphor, Eucalyptol, Menthol, Potassium Phosphate, Purified Water, Sodium Chloride, Sodium Phosphate, Tyloxapol, Benzalkonium Chloride, Chlorhexidine Gluconate, Edetate Disodium.

VICKS, VAPORUB

Manufacturer: Procter & Gamble Company

Active Ingredient & Concentration:
2.6% Menthol + 4.7% Camphor
+ 1.2% Eucalyptus Oil

Product Form:
Ointment

Other Ingredients:
Cedarleaf Oil, Mineral Oil, Nutmeg Oil, Petrolatum, Spirits of Turpentine, Thymol.

VICKS, VATRONOL

Manufacturer: Procter & Gamble Company

Active Ingredients & Concentration:
0.5% Ephedrine Sulfate

Product Form:
Liquid

Other Ingredients:
Camphor, Cedarleaf Oil, Eucalyptol, Menthol, Nutmeg Oil, Potassium Phosphate, Purified Water, Sodium Chloride, Sodium Phosphate, Tyloxapol, Thimerosal.

PART II

SUPPLIER'S FORMULATIONS

PSEUDOEPHEDRINE HYDROCHLORIDE CAPSULE I

Supplier: Aqualon Company

Ingredient:	Wt.%
Pseudoephedrine Hydrochloride	24.00
Hydroxyethylcellulose,NF (**Natrosol 250HX**)	15.00
Lactose, Anhydrous	60.00
Magnesium Stearate	1.00
	100.00

Mixing Procedure:

Blend all the ingredients in a twin-shell blender for 10 minutes. Fill No. 0 gelatin capsules using a Zanasi LZ64 capsule filling machine. Fill weight is 500mg. The tamping force on the dosator piston is 200 Newtons.

PSEUDOEPHEDRINE HYDROCHLORIDE CAPSULE II

Supplier: Aqualon Company

Ingredient:	Wt.%
Pseudoephedrine Hydrochloride	24.00
Hydroxypropylcellulose (**Klucel HXF**)	25.00
Lactose, Anhydrous	50.00
Magnesium Stearate	1.00
	100.00

Mixing Procedure:

Blend all the ingredients in a twin-shell blender for 10 minutes. Fill No. 0 gelatin capsules using a Zanasi LZ64 capsule filling machine. Fill weight is 500mg. The tamping force on the dosator piston is 200 Newtons.

PHENYLPROPANOLAMINE TABLETS

Supplier: Aqualon Company

	Ingredient:	Wt.%
A	Phenylpropanolamine, NF	15.00
B	Hydroxypropylcellulose (**Klucel HXF**)	30.00
C	Microcrystalline Cellulose, NF	51.60
D	Povidone, USP	2.40
E	Magnesium Stearate	1.00
		100.00

Mixing Procedure:

Mix Parts A, B and a portion of C in a low-shear planetary mixer. Dissolve Part D in water and add the solution to the mixer. The mixture is then granulated, dried and milled. This mixture along with the remaining Part C are mixed in "V" blender. Part E is added and blended with the other ingredients. Tablets are prepared on a rotary tablet press with capsule-shaped tooling.

TERFENADINE SUSPENSION

Supplier: BASF Corp.

	Ingredient:	Wt.%
A	Terfenadine	1.20
B	Poloxamer 407 (**Lutrol F 127**)	3.00
C	PEG-40 Hydrogenated Castor Oil (**Cremphor RH 40**)	3.60
D	Water	92.20
		100.00

Mixing Procedure:

Dissolve Parts B and C in Part D at 40°C. Slowly add Part A to the solution with stirring.

THEOPHYLLINE + EPHEDRINE TABLETS

Supplier: BASF Corp.

Ingredient:	Wt.%
Theophylline	43.46
Ephedrine Hydrochloride	5.02
Lactose(and)Povidone(and)Crospovidone (**Ludipress**)	50.20
Colloidal Silicon Dioxide	0.66
Magnesium Stearate	0.66
	100.00

Mixing Procedure:

Mix all the ingredients, pass through a sieve and press at very low compressive force. Tablet weight = 302mg, diameter = 8mm.

CHLORPHENIRAMINE MALEATE (4mg) + PSEUDOEPHEDRINE HYDROCHLORIDE (60mg) TABLET

Supplier: FMC Corp.

	Ingredient:	Wt.%
A	Chlorpheniramine Maleate Powder	1.82
B	Pseudoephedrine Hydrochloride Powder	27.27
C	Microcrystalline Cellulose (**Avicel® PH-101**)	16.95
D	Lactose, Anhydrous	51.36
E	Croscarmellose Sodium (**Ac-Di-Sol® SD-711**)	1.00
F	Fumed Silica	0.50
G	Stearic Acid	0.60
H	Magnesium Stearate, Fine Powder	0.50
		100.00

Mixing Procedure:

Screen all ingredients through a 20 mesh sieve. Blend Parts A, B and C in V-shaped blender for 3 minutes. Add Parts D, E and F to the batch and blend for 10 minutes. Add Part G to the batch and blend for 3 minutes. Add Part H to the batch and blend for 5 minutes. Compress using 5/16" standard concave punches.

DIPHENHYDRAMINE HYDROCHLORIDE TABLET (25mg)

Supplier: FMC Corp.

	Ingredient:	Wt.%
A	Diphenhydramine Hydrochloride Powder	10.00
B	Microcrystalline Cellulose (**Avicel PH-200**)	20.00
C	Lactose	68.20
D	Croscarmellose Sodium (**Ac-Di-Sol SD-711**)	1.00
E	Fumed Silica	0.30
F	Stearic Acid XXX*	0.20
G	Magnesium Stearate, Fine Powder	0.30
		100.00

Mixing Procedure:

Screen all the ingredients through a 20 mesh sieve. Blend Parts A, B, D and E in a V-shaped blender for 3 minutes. Add Part C to the batch and blend for 10 minutes. Add Part F to the batch and blend for 3 minutes. Add Part G to the batch and blend for 5 minutes. Tablet using 3/8" FBE tooling.

THEOPHYLLINE (130mg) + EPHEDRINE SULFATE (24mg) TABLET

Supplier: FMC Corp.

	Ingredient:	Wt.%
A	Theophylline, Anhydrous	40.63
B	Ephedrine Sulfate Powder	7.50
C	Microcrystalline Cellulose (**Avicel PH-101**)	16.25
D	Lactose, Anhydrous	32.87
E	Croscarmellose Sodium (**Ac-Di-Sol SD-711**)	1.00
F	Fumed Silicon Dioxide	0.50
G	Stearic Acid XXX*	0.75
H	Magnesium Stearate, Fine Powder	0.50
		100.00

Mixing Procedure:

Screen Parts A, B, G and H, separately, through a 40 mesh sieve. Blend Parts B, C and E in a "V" blender for 3 minutes. Add Part A to the batch and blend 3 minutes. Add Parts D and F to the batch and blend 15 minutes. Add Part G to the batch and blend 3 minutes. Add Part H to the batch and blend 5 minutes. Tablet on Manesty Express 20 machine, using 3/8" flat bevel punches, to a hardness of 7.0kg. Average tablet weight is 319mg.

---

* XXX is a commonly used abbreviation for triple-pressed Stearic Acid

CHILDREN'S DECONGESTANT COLD TABLETS

Supplier: International Specialty Products

	Ingredient:	Wt.%
A	Acetylsalicylic Acid, USP	6.14
B	Phenylpropanolamine Hydrochloride	0.46
C	Flavor	1.51
D	Color	0.15
E	Sodium Bicarbonate	41.70
F	Citric Acid, Anhydrous Granular	24.64
G	Citric Acid, Anhydrous Powder	24.64
H	Polyvinyl Pyrrolidone, 10% Aqueous Solution (**Plasdone® K-30**)	0.76
		100.00

Mixing Procedure:

Convert Part E to 7-9% Sodium Carbonate by heating in a forced-air draft oven at 100°C for 45 minutes. Stir twice, at 15 minute intervals. Cool and mix with Parts F and G in a planetary mixer for 10 minutes. Quickly add Part H until the mass is merely damp and the water is uniformly dispersed. A mild reaction occurs. Immediately transfer to paper-lined drying trays and spread evenly. Place in a forced-air draft oven at 70°C for 20 minutes. Cool and granulate through a 12 mesh screen. Mix ingredients B, C and D. Combine this mixture, the granulated mixture and Part A in tumble blender until uniform. Compress to 5/8" diameter, flat-faced, beveled-edge tablets weighing 1.32gm each. Stabilize the tablets by heating in a forced-air draft oven for 1 hour at 60°C and package in air-tight aluminum foil.

CHAPTER 11

CONTRACEPTIVE PRODUCTS

This chapter contains formulations for spermicidal products. The active spermicide ingredients that are most frequently used in brand name products are Nonoxynol-9 or Octoxynol-9.

Product forms include gels, suppositories (also called vaginal inserts) and foams.

PART I

BRAND NAME FORMULATIONS

CONCEPTROL

Manufacturer: Ortho Pharmaceutical Corp.

Active Ingredients & Concentration:
Gel: 4.00% Nonoxynol-9
Suppository: 150mg Nonoxynol-9/ Suppository

Product Forms:
Gel
Suppository

Other Ingredients:
Gel contains Lactic Acid, Methylparaben, Povidone, Propylene Glycol, Purified Water, Carboxymethylcellulose Sodium, Sorbic Acid, Sorbitol.
Inserts contain Lauroamphodiacetate Sodium Trideceth Sulfate, Polyethylene Glycol 1000, Polyethylene Glycol 1450, Povidone.

DELFIN

Manufacturer: Ortho Pharmaceutical Corp.

Active Ingredients & Concentration:	Product Form:
12.5% Nonoxynol-9	Foam

Other Ingredients:
Benzoic Acid, Cellulose Gum, Glacial Acetic Acid, Cetyl Alcohol, Fragrance, Methylparaben, Phosphoric Acid, Polyvinyl Alcohol, Propylene Glycol, Purified Water, Stearamidoethyl Diethylamine, Sorbic Acid, Stearic Acid, A-31 Propellant.

ENCARE

Manufacturer: Thompson Medical Company

Active Ingredient & Concentration:	Product Form:
100 mg Nonoxynol-9/Suppository	Suppository

Other Ingredients:
Polyethylene Glycols, Sodium Bicarbonate, Sodium Citrate, Tartaric Acid.

GYNOL II

Manufacturer: Ortho Pharmaceutical Corp.

Active Ingredient & Concentration:	Product Form:
2.0% Nonoxynol-9	Gel

Other Ingredients:
Lactic Acid, Methylparaben, Povidone, Propylene Glycol, Purified Water, Carboxymethylcellulose Sodium, Sorbic Acid, Sorbitol.

GYNOL II, EXTRA STRENGTH

Manufacturer: Ortho Pharmaceutical Corp.

Active Ingredient & Concentration: 3.0% Nonoxynol-9

Product Form: Gel

Other Ingredients:
Lactic Acid, Methylparaben, Povidone, Propylene Glycol, Purified Water, Carboxymethylcellulose Sodium, Sorbic Acid, Sorbitol.

ORTHO-GYNOL

Manufacturer: Ortho Pharmaceutical Corp.

Active Ingredient & Concentration: 1.0% Octoxynol-9

Product Form: Gel

Other Ingredients:
Benzoic Acid, Castor Oil, Fragrance, Glacial Acetic Acid, Methylparaben, Sorbic Acid, Potassium Hydroxide, Propylene Glycol, Purified Water, Carboxymethylcellulose Sodium.

SEMICID

Manufacturer: Whitehall Laboratories, Inc.

Active Ingredient & Concentration: 100 mg Nonoxynol-9/Suppository

Dosage Form: Suppository

Other Ingredients:
Benzethonium Chloride, Citric Acid, D&C Red No. 21 Lake, Methylparaben, Polyethylene Glycol, Water.

PART II

SUPPLIER'S FORMULATIONS

CONTRACEPTIVE VAGINAL GEL

Supplier: Gattefossé s. a.

	Ingredient:	Wt.%
A	PEG-6(and)PEG-32(and)Glycol Stearate (**Tefose 63**)	5.00
	Apricot Kernel Oil PEG-6 Esters (**Labrafil M 1944 CS**)	3.00
B	Demineralized Water	81.60
	Hydroxyethylcellulose	8.00
C	Benzalkonium Chloride, 50 wt% in Water	2.40
		100.00

Mixing Procedure:

Prepare Part B at room temperature. Heat Part B to 75°C and add Part A ingredients while stirring. Cool with gentle stirring, to 30°C. Add Part C and stir.

CONTRACEPTIVE VAGINAL SUPPOSITORIES

Supplier: Gattefossé s. a.

Ingredient:	Wt.%
Benzalkonium Chloride	1.18
Hemi-Synthetic Glycerides with Hydrophilic Agent (**Ovucire WL 3264**)	98.82
	100.00

Mixing Procedure:

Melt the **Ovucire WL 3264** at 50°C. Allow to cool to 40°C and add the Benzalkonium Chloride while mixing with a turbine mixer. Continue cooling to 33°C, then pour into molds previously chilled to 0 to -5°C. Scrape after 12 minutes and remove from the mold after 18 minutes.

CHAPTER 12

DANDRUFF, SEBORRHEA & PSORIASIS TREATMENTS

Zinc Pyrithione, Selenium Sulfide and Coal Tar are the three primary active ingredients used in OTC dandruff, seborrhea and psoriasis treatments. These actives are delivered to the scalp in the form of shampoos, creams or lotions. Hydrocortisone creams and lotions (presented in Chapter 8 of this book) are sometimes used to relieve the itching associated with these scalp problems.

PART I

BRAND NAME FORMULATIONS

DENOREX

Manufacturer: Whitehall Laboratories, Inc.

Active Ingredient & Concentration: 1.8% Coal Tar

Product Form: Shampoo

Other Ingredients:
Alcohol (7.5% by volume), TEA-Lauryl Sulfate, Chloroxylenol, Lauramide DEA, Menthol, Stearic Acid, Water, Hydroxypropyl Methylcellulose, Fragrance.

DENOREX, WITH CONDITIONER

Manufacturer: Whitehall Laboratories, Inc.

Active Ingredient & Concentration: 1.8% Coal Tar

Product Form: Shampoo

Other Ingredients:
Alcohol (7.5% by volume), TEA-Lauryl Sulfate, Chloroxylenol, Citric Acid, Hydroxypropyl Methylcellulose, Fragrance, Lauramide DEA, Menthol, PEG-27 Lanolin, Polyquaternium-11, Water.

DENOREX, EXTRA STRENGTH

Manufacturer: Whitehall Laboratories, Inc.

Active Ingredient & Concentration: 2.5% Coal Tar

Product Form: Shampoo

Other Ingredients:
Alcohol (10.4% by volume), TEA-Lauryl Sulfate, FD&C Red No. 40, Hydroxypropyl Methylcellulose, Glycol Distearate, Fragrance, Chloroxylenol, Lauramide DEA, Menthol, Water.

DENOREX, EXTRA STRENGTH WITH CONDITIONER

Manufacturer: Whitehall Laboratories, Inc.

Active Ingredient & Concentration: 2.5% Coal Tar

Product Form: Shampoo

Other Ingredients:
Alcohol (10.4% by volume), TEA-Lauryl Sulfate, FD&C Red No. 40, Hydroxypropyl Methylcellulose, Glycol, Distearate, Cocodimonium Hydrolyzed Protein, Fragrance, Chloroxylenol, Lauramide DEA, Menthol, PEG-27 Lanolin, Polyquaternium-6, Water.

HEAD & SHOULDERS

Manufacturer: Procter & Gamble Company

Active Ingredient & Concentration: 1.0% Zinc Pyrithione

Product Forms: Shampoo, Cream

Other Ingredients:
Normal to oily hair cream contains Water, Sodium Cocoglyceryl Ether Sulfonate, Sodium Chloride, Sodium Lauroyl Sarcosinate, Lauramide DEA, Cocoyl Sarcosine, Fragrance, FD&C Blue No. 1.
Normal to dry hair shampoo contains Water, Ammonium Laureth Sulfate, Ammonium Lauryl Sulfate, Cocamide MEA, Dimethicone, Glycol Distearate, Ammonium Xylenesulfonate, Cetrimonium Chloride, Cetyl Alcohol, Stearyl Alcohol, DMDM Hydantoin, Sodium Chloride, FD&C Blue No. 1.

HEAD & SHOULDERS, 2-IN-1 WITH CONDITIONER

Manufacturer: Procter & Gamble Company

Active Ingredient & Concentration: 1.0% Zinc Pyrithione

Product Form: Shampoo

Other Ingredients:
Water, Ammonium Laureth Sulfate, Ammonium Lauryl Sulfate, Cocamide MEA, Dimethicone, Glycol Distearate, Ammonium Xylenesulfonate, Cetrimonium Chloride, Cetyl Alcohol, Sodium Phosphate, Disodium Phosphate, Stearyl Alcohol, DMDM Hydantoin, Sodium Chloride, FD&C Blue No. 1, Fragrance.

HEAD & SHOULDERS, DRY SCALP FORMULAS

Manufacturer: Procter & Gamble Company

Active Ingredient & Concentration: 1.0% Zinc Pyrithione

Product Form: Shampoo

Other Ingredients:
Regular, Conditioning and 2-In-1 formulae all contain Water, Ammonium Laureth Sulfate, Ammonium Lauryl Sulfate, Cocamide MEA, Dimethicone, Glycol Distearate, Ammonium Xylenesulfonate, Cetrimonium Chloride, Cetyl Alcohol, Stearyl Alcohol, DMDM Hydantoin, Sodium Chloride, FD&C Blue No. 1, Fragrance

---

HEAD & SHOULDERS, INTENSIVE TREATMENT FORMULA

Manufacturer: Procter & Gamble Company

Active Ingredient & Concentration: 1.0% Selenium Sulfide

Product Form: Shampoo

Other Ingredients:
Regular, and 2-In-1 formulae contain Water, Ammonium Laureth Sulfate, Ammonium Lauryl Sulfate, Cocamide MEA, Dimethicone, Glycol Distearate, Ammonium Xylenesulfonate, Cetrimonium Chloride, Cetyl Alcohol, Stearyl Alcohol, DMDM Hydantoin, Sodium Chloride, Hydroxypropyl Methylcellulose, FD&C Red No. 4, Fragrance

---

NEUTROGENA, T/GEL

Manufacturer: Neutrogena Corp.

Active Ingredient & Concentration: 0.5% Coal Tar

Product Form: Shampoo

Other Ingredients:
Purified Water, Sodium Laureth Sulfate, Cocamide DEA, Laureth-4, Fragrance, Cocamidopropyl Betaine, Imidazolidinyl Urea, Propylparaben, Methylparaben, Citric Acid, Sodium Chloride, Tetrasodium EDTA.

NEUTROGENA, T/SAL

Manufacturer: Neutrogena Corp.

Active Ingredient & Concentration: 2% Salicylic Acid

Product Form: Shampoo

Other Ingredients:
Purified Water, Sodium C14-16 Olefin Sulfonate, Cocamidopropyl Betaine, Isostearylamidopropyl PG-Dimonium Chloride, Hexylene Glycol, Sodium C12-15 Pareth-15 Sulfonate, Glycerin, Fragrance, Colors.

SELSUN BLUE, REGULAR HAIR FORMULA

Manufacturer: Ross Laboratories

Active Ingredient & Concentration: 1% Selenium Sulfide

Product Form: Shampoo

Other Ingredients:
Ammonium Laureth Sulfate, Ammonium Lauryl Sulfate, Citric Acid, Cocamide DEA, Cocamidopropyl Betaine, DMDM Hydantoin, FD&C Blue No. 1, Fragrance, Hydroxypropyl Methylcellulose, Magnesium Aluminum Silicate, Purified Water, Sodium Chloride, Titanium Dioxide.

SELSUN BLUE, DRY HAIR FORMULA

Manufacturer: Ross Laboratories

Active Ingredient & Concentration:
1% Selenium Sulfide

Product Form:
Shampoo

Other Ingredients:
Acetylated Lanolin Alcohol, Ammonium Laureth Sulfate, Ammonium Lauryl Sulfate, Cetyl Acetate, Citric Acid Cocamide DEA, Cocamidopropyl Betaine, DMDM Hydantoin, FD&C Blue No. 1, Fragrance, Hydroxypropyl Methylcellulose, Magnesium Aluminum Silicate, Polysorbate 80, Purified Water, Sodium Chloride, Titanium Dioxide.

SELSUN BLUE, OILY HAIR FORMULA

Manufacturer: Ross Laboratories

Active Ingredient & Concentration:
1% Selenium Sulfide

Product Form:
Shampoo

Other Ingredients:
Ammonium Laureth Sulfate, Ammonium Lauryl Sulfate, Citric Acid, Cocamide DEA, Cocamidopropyl Betaine, DMDM Hydantoin, FD&C Blue No. 1, Fragrance, Hydroxypropyl Methylcellulose, Magnesium Aluminum Silicate, Purified Water, Sodium Chloride, Titanium Dioxide.

SELSUN BLUE, EXTRA CONDITIONING FORMULA

Manufacturer: Ross Laboratories

Active Ingredient & Concentration:
1% Selenium Sulfide

Product Form:
Shampoo

Other Ingredients:
Aloe, Ammonium Laureth Sulfate, Ammonium Lauryl Sulfate, Citric Acid, Cocamide DEA, Ditallow Phthalic Acid Amide, Dimethicone, DMDM Hydantoin, FD&C Blue No. 1, Fragrance, Hydroxypropyl Methylcellulose, Purified Water, Sodium Citrate, Sodium Isostearoyl Lactylate, Titanium Dioxide.

SELSUN BLUE, MEDICATED TREATMENT FORMULA

Manufacturer: Ross Laboratories

Active Ingredient & Concentration: 1% Selenium Sulfide

Product Form: Shampoo

Other Ingredients:
Ammonium Laureth Sulfate, Ammonium Lauryl Sulfate, Citric Acid, Cocamide DEA, Cocamidopropyl Betaine, DMDM Hydantoin, D&C Red No. 33, FD&C Blue No. 1, Fragrance, Hydroxypropyl Methylcellulose, Magnesium Aluminum Silicate, Menthol, Purified Water, Sodium Chloride, TEA-Lauryl Sulfate.

SELSUN GOLD, FOR WOMEN

Manufacturer: Ross Laboratories

Active Ingredient & Concentration: 1% Selenium Sulfide

Product Form: Shampoo

Other Ingredients:
Ammonium Laureth Sulfate, Ammonium Lauryl Sulfate, Citric Acid, Cocamide DEA, Ditallow Phthalic Acid Amide, Dimethicone, DMDM Hydantoin, Fragrance Hydroxypropyl Methylcellulose, Purified Water, Sodium Citrate.

TEGRIN

Manufacturer: Block Drug Company

Active Ingredient & Concentration: 1.1% Coal Tar

Product Form: Shampoo

Other Ingredients:
Water, Sodium Lauryl Sulfate, Ammonium Lauryl Sulfate, Alcohol (7.0%), Glycol Stearate(and)Sodium Laureth Sulfate(and)Hexylene Glycol, Lauramide DEA, Fragrance, Hydroxypropyl Methylcellulose, Citric Acid, Guar Hydroxypropyltrimonium Chloride, Methylparaben, Propylparaben, FD&C Blue No. 1.

X-SEB

Manufacturer: Baker Cummins Dermatologicals, Inc.

Active Ingredient & Concentration:	Product Form:
1% Zinc Pyrithione	Shampoo

Other Ingredients:
Purified Water, Ammonium Lauryl Sulfate, Ammonium Laureth Sulfate, Glycol Distearate, Lauramide DEA Salicylic Acid, PEG 75 Lanolin, Ammonium Xylene Sulfate, Triethanolamine, Menthol, Fragrance, Methylchloroisothiazolinone, FD&C Blue No. 1 , Methylisothiazolinone.

X-SEB, PLUS

Manufacturer: Baker Cummins Dermatologicals, Inc.

Active Ingredient & Concentration:	Product Form:
1% Zinc Pyrithione	Shampoo

Other Ingredients:
Purified Water, Ammonium Lauryl Sulfate, Ammonium Laureth Sulfate, Glycol Distearate, Lauramide DEA, Salicylic Acid, PEG 75 Lanolin, Polyquaternium-7, Ammonium Xylene Sulfate, Triethanolamine, Menthol, Fragrance, FD&C Blue No. 1, Methylchloroisothiazolinone, Methylisothiazolinone.

X-SEB T

Manufacturer: Baker Cummins Dermatologicals, Inc.

Active Ingredient & Concentration:	Product Form:
2% Coal Tar	Shampoo

Other Ingredients:
Purified Water, TEA-Lauryl Sulfate, Ethylene Glycol Distearate, Lauramide DEA, Salicylic Acid, PEG 75 Lanolin, Menthol, Chloroxylenol, Edetate Disodium, Fragrance, FD&C Blue No. 1, Hydroxypropyl Methylcellulose.

X-SEB T, PLUS

Manufacturer: Baker Cummins Dermatologicals, Inc.

Active Ingredient & Concentration:	Product Form:
2% Coal Tar	Shampoo

Other Ingredients:
Purified Water, TEA-Lauryl Sulfate, Glycol Distearate, Lauramide DEA, Salicylic Acid, PEG 75 Lanolin, Polyquaternium-7, Menthol, Chloroxylenol, Edetate Disodium, Fragrance, FD&C Blue No. 1, Hydroxypropyl Methylcellulose.

ZINCON

Manufacturer: Lederle Laboratories

Active Ingredient & Concentration:	Product Form:
1% Zinc Pyrithione	Shampoo

Other Ingredients:
Water, Sodium Methyl Cocoyl Taurate, Cocamide DEA, Sodium Chloride, Magnesium Aluminum Silicate, Sodium Cocoyl Isethionate, Fragrance, Glutaraldehyde, D&C Green No. 5, Citric Acid or Sodium Hydroxide, if necessary to adjust pH.

PART II

SUPPLIER'S FORMULATIONS

DANDRUFF CONTROL SHAMPOO

Supplier: American Colloid Company

	Ingredient:	Wt.%
A	Water	42.00
B	Purified Bentonite (**Polargel NF**)	1.00
C	Sodium Lauryl Sulfate, 30%	50.00
	Glycol Stearate	2.70
	Lauramide DEA	2.70
D	Polysorbate 80(and)Sulfur	1.00
	Sodium Chloride	0.60
	Fragrance, Color, Preservative	q.s.
		100.00

Mixing Procedure:

Heat Part A to 75-80°C. Sprinkle Part B into Part A with agitation. When Part B is well dispersed, add the Part C ingredients and mix well. Keep the temperature constant. When Parts A+B+C are completely uniform, cool and continue stirring. At 40°C, add the Part D ingredients, one at a time, with stirring. Adjust the pH to 7.5-8.5 with 50% citric acid solution. Continue stirring until cool.

ANTIDANDRUFF SHAMPOO

Supplier: Aqualon Company

	Ingredient:	Wt.%
A	Water	54.75
B	Hydroxyethylcellulose (**Natrosol 250HR**)	0.75
C	TEA-Lauryl Sulfate	34.70
D	PPG-20 Lanolin Alcohol Ether	4.30
E	Glycol Stearate	2.00
F	Cocamide MEA	1.50
G	Zinc Pyrithione, 48%	2.00
H	Fragrance, Preservative	q.s.
		100.00

Mixing Procedure:

Add the **Natrosol 250HR** to the Water and mix. In a separate vessel, combine Parts C, D and E, heat to 80°C and mix. Cool to 50°C. Add Parts F and G to this part of the batch and mix. Add this mixture to the **Natrosol 250HR** solution. Cool to 40°C and add Part H.

ANTIDANDRUFF SHAMPOO

Supplier: Calgon Corp.

	Ingredient:	Wt.%
	Xanthan Gum (**Keltrol CG**)	1.00
A	Glycerin	5.00
	Deionized Water	54.80
	TEA-Lauryl Sulfate	25.00
	Sodium Lauroyl Sarcosinate	10.00
C	Zinc Pyrithione, 48%	4.20
	Fragrance, Preservative	q.s.
		100.00

Mixing Procedure:

Slurry the **Keltrol CG** in the Glycerin and add the slurry to the Water with vigorous agitation. Mix until the **Keltrol CG** is fully hydrated. Add the Part B ingredients to Part A. Homogenize.

ANTIDANDRUFF SHAMPOO I

Supplier: Croda, Inc.

	Ingredient:	Wt.%
A	Cocodimonium Hydroxypropyl Hydrolyzed Collagen (**Croquat M**)	1.00
B	TEA-Lauryl Sulfate	20.00
C	Cocamide MEA (**Incromide CME**)	4.00
D	Zinc Pyrithione, 48%	2.00
E	Glycol Distearate	7.00
F	Sodium Chloride	0.50
G	Propylene Glycol	0.50
H	Propylparaben	0.10
I	Methylparaben	0.30
J	Deionized Water	64.60
K	Citric Acid to ph 4.7	q.s.
L	Color	q.s.
		100.00

Mixing Procedure:

Mix Parts D and J, heat to 70°C and stir for 5 minutes. Add Parts F, G, H and I. When the mixture has dissolved, add Parts B, C and E. Heat with mixing to 75°C. Cool, with mixing, to 40°C and add Part A. Adjust the pH with Part K and add Part L, with mixing, until uniform.

ANTIDANDRUFF SHAMPOO II

Supplier: Croda, Inc.

	Ingredient:	Wt.%
A	Deionized Water	61.50
	Magnesium Aluminum Silicate	1.00
B	Xanthan Gum	0.50
	Propylene Glycol	3.50
C	Ammonium Lauryl Sulfate	20.00
	Cocamide DEA (**Incromide CA**)	3.00
D	Cocamidodipropyl Hydroxysultaine (**Crosultaine C-50**)	7.00
	Polyol Alkoxy Ester (**Crothix**)	0.50
	Zinc Pyrithione, 48%	2.00
E	Propylene Glycol(and)Diazolidinyl Urea(and) Propylparaben(and)Methylparaben	1.00
		100.00

Mixing Procedure:

Heat Part A to 80-85°C. Combine Part B ingredients and add to Part A using high shear. Add Part C with mixing. With gentle mixing, combine the Part D ingredients and heat to 80-85°C. Add Part D to Parts A+B+C with mixing. Cool to 30-35°C with mixing and add the Part E ingredients.

ANTIDANDRUFF SHAMPOO

Supplier: Gattefossé s.a.

	Ingredient:	Wt.%
	Demineralized Water	53.95
	Disodium Laureth Sulfosuccinate	15.00
A	Sodium Laureth Sulfate	15.00
	Cocamidopropyl Betaine	7.00
	Propylene Glycol(and)Water(and)Nettle Extract (**Vegetol Common Nettle MCF 801 Hydro**)	5.00
	Cocamide DEA	3.00
B	Piroctone Olamine	0.75
	Preservative	q.s.
	Fragrance	0.10
C	Octoxynol-11(and)Polysorbate 20 (**Solubilisant Gamma 2420**)	0.20
D	Citric Acid	q.s.
	Sodium Chloride	q.s.
		100.00

Mixing Procedure:

Mix the Part A ingredients at room temperature in the order shown. Mix and gently heat the Part B ingredients until the Piroctone Olamine has melted. Add Part B to Part A. Mix the Part C ingredients and add to Parts A+B with mixing. Adjust the pH to 5.8-6.0 with Citric Acid and adjust the viscosity, if necessary, with Sodium Chloride.

ANTI-SEBORRHEA CLEAR LOTION

Supplier: Gattefossé s.a.

	Ingredient:	Wt.%
A	Progesterone	0.50
B	Ethanol	10.00
C	Ethoxydiglycol (**Transcutol**)	50.00
D	Demineralized Water	38.50
E	Propylene Glycol(and)Sodium Methylparaben(and) Sodium Dehydroacetate(and)Sorbic Acid (and) Tetrasodium EDTA (**Fondix G Bis**)	1.00
		100.00

Mixing Procedure:

Dissolve Part A in Part B at ambient temperature. Add Part C, D and E in the order shown, with mixing.

ANTI-SEBORRHEA CREAM I

Supplier: Gattefossé s.a.

	Ingredient:	Wt.%
	Cetyl Alcohol(and)Ceteth-20(and)Steareth-20 (**Emulcire 61 WL 2659**)	15.00
A	Hydrogenated Palm/Palm Kernel PEG-6 Esters (**Labrafil M 2130 CS**)	3.00
	Mineral Oil	8.00
	Sorbic Acid	0.05
	Sodium Methylparaben	0.05
	Demineralized Water	72.90
B	Anthralin	1.00
		100.00

Mixing Procedure:

Mix and heat the Part A ingredients to 75°C. Cool and add Part B, with stirring. Homogenize, if necessary.

ANTI-SEBORRHEA CREAM II

Supplier: Gattefossé s.a.

	Ingredient:	Wt.%
	Cetyl Alcohol(and)Ceteth-20(and)Steareth-20 (**Emulcire 61 WL 2659**)	15.00
A	Hydrogenated Palm/Palm Kernel PEG-6 Esters (**Labrafil M 2130 CS**)	3.00
	Octyldodecyl Myristate (**MOD**)	10.00
	Sodium Methylparaben	0.05
B	Sorbic Acid	0.05
	Demineralized Water	61.76
	Coal Tar	5.00
C	Allantoin	0.10
	Hydrolyzed Soy Protein (**Phylderm Vegetal**)	5.00
D	α-Tocopherol Acetate	0.02
	Vitamin A	0.02
		100.00

Mixing Procedure:

Mix and heat the Part A ingredients to 80°C. Cool with rapid stirring. At 30°C, add the Part B ingredients. Add the Part C ingredients to Parts A+B, then add Part D ingredients to Parts A+B+C. Homogenize, if necessary.

ANTIDANDRUFF SHAMPOO

Supplier: Hüls America, Inc.

	Ingredient:	Wt.%
	Sodium Laureth Sulfate (**Marlinat 242/28**)	35.00
	Cocamidopropyl Betaine (**Ampholyt JB 130/K**)	15.00
A	PEG-6 Caprylic/Capric Glycerides (**Softigen 767**)	4.00
	Talloweth-60 Myristol Glycol	
	Demineralized Water	
	Ethoxydiglycol(and)Propylene Glycol(and) Nettle Extract(and)Butylene Glycol	1.50
B	Piroctone Olamine	1.00
	d-Panthenol	0.20
	Fragrance, Color, Preservative	q.s.
		100.00

Mixing Procedure:
Mix the Part A ingredients in the order shown. Warm while stirring until homogeneous. Add the Part B ingredients at approximately 30°C. Adjust the pH to 5.5-6.5.

ANTIDANDRUFF SHAMPOO

Supplier: International Specialty Products

	Ingredient:	Wt.%
A	Deionized Water	56.73
B	PMV/MA Decadiene Crosspolymer(**Stabileze® 06**)	1.25
C	Ammonium Lauryl Sulfate	25.00
D	Ammonium Laureth Sulfate	10.00
E	Lauryl Pyrrolidone (**Sulfadone® LP-300**)	1.50
F	Sodium Hydroxymethylglycinate (**Suttocide® A**)	3.00
G	Zinc Pyrithione	2.00
H	Fragrance	0.25
I	Colors	0.27
		100.00

Mixing Procedure:
Add Part B to Part A and mix well. Heat to 80°C and hold at that temperature for 40-50 minutes with agitation. Cool to 70°C. Add Part C slowly while mixing. Add Part D at 70°C while mixing. Add Part E at 70°C and mix well. Cool the batch to 40°C and add Part F. Mix well. Add Part G slowly with constant agitation. Cool to room temperature and add Parts H and I. Mix until uniform.

ANTIDANDRUFF SHAMPOO

Supplier: Laboratoires Serobiologiques

	Ingredient:	Wt.%
	Distilled Water	58.60
	Hexamidine Diisethionate (**Elestab HP-100**)	0.10
A	Preservative	2.00
	PEG-120 Methyl Glucose Dioleate	1.00
	Sodium Laureth Sulfate	25.00
	Polyquaternium-11	2.00
B	Potassium Undecylenoyl Hydrolyzed Collagen	2.00
	Cocamidopropyl Betaine	8.00
	Oleamide DEA	1.30
		100.00

Mixing Procedure:

Mix and heat all the Part A components to 50°C. In a separate vessel, mix the Part B ingredients at room temperature. Add Part B to Part A while stirring. Cool to room temperature.

CONDITIONING, ANTIDANDRUFF SHAMPOO

Supplier: Rhône-Poulenc, Inc.

	Ingredient:	Wt.%
A	Guar Hydroxypropyltrimonium Chloride (**Jaguar C-17**)	0.50
B	Water	35.00
C	Xanthan Gum (**Rhodigel EZ**)	0.05
D	Citric Acid to pH 5-7	q.s.
E	Sodium Lauryl Sulfate (**Rhodapon SB 8208S**)	45.00
F	Cocamide MEA (**Alkamide C-212**)	1.80
G	Glycol Stearate (**Alkamuls EGMS**)	0.90
H	Zinc Pyrithione, 48%	2.00
I	Water	5.00
J	Citric Acid to pH 5.5-6.5	q.s.
K	Fragrance, Color, Preservative	q.s.
L	Cocamidopropyl Betaine (**Mirataine CBC**)	6-14.00

Mixing Procedure:

With rapid but smooth agitation, disperse Part A in Part B. Heat to 40-45°C with smooth agitation. Disperse Part C in Parts A+B. With vigorous agitation, heat to 60°C. Adjust pH with Part D. With smooth agitation, blend in Part E. Heat, with mixing, to 70-75°C. With smooth agitation, blend in Parts F and G. Mix until uniform. Combine Parts H and I and mix until uniform. Add Parts H+I to the batch slowly and mix until uniform. Cool to 40-45°C and adjust pH with Part J. Add Part K and mix until uniform. Adjust the final formula viscosity to 7,00-11,000 cps. with Part L.

ZINC PYRITHIONE SHAMPOO I

Supplier: Rohm and Haas Company

Ingredient:	Wt.%
Deionized Water	66.58
Aculyn 33	6.67
Sodium Hydroxide, 50%	0.75
Decyl Glucoside	13.50
Cocamidopropyl Dimethylaminohydroxypropyl Hydrolyzed Collagen	8.50
PEG-200 Glyceryl Stearate	3.00
Zinc Pyrithione, 48%	1.00
Preservative	q.s.
	100.00

Mixing Procedure:

Add the ingredients in the order shown (very important). Adjust ph to 7.

ZINC PYRITHIONE SHAMPOO II

Supplier: Rohm and Haas Company

	Ingredient:	Wt.%
	Deionized Water	45.98
	Preservative	q.s.
A	**Aculyn 22**	6.70
	Cocamidopropyl Betaine	10.00
	Ammonium Lauryl Sulfate	30.00
	Sodium Hydroxide, 10%	1.90
B	Glycolic Acid	1.22
	Zinc Pyrithione, 48%	4.20
	Color	q.s.
		100.00

Mixing Procedure:

Mix the Part A ingredients in the order shown. Add Sodium Hydroxide and stir for 15 minutes. Add remaining Part B ingredients in the order shown. For optimum stability, keep ph <4.2.

ANTIDANDRUFF SHAMPOO

Supplier: Southern Clay Products Company

	Ingredient:	Wt.%
	Montmorillonite (**Gelwhite H, NF**)	1.00
A	Hydroxypropyl Methylcellulose	0.80
	Water	45.80
B	Zinc Pyrithione, 48%	4.20
C	Cocamide MEA	5.00
D	TEA-Lauryl Sulfate	40.00
	Triethanolamine	3.20
	Color, Fragrance, Preservatives	q.s.
		100.00

Mixing Procedure:

Add the **Gelwhite H, NF** to the water, agitating at maximum available shear until smooth. Heat to 50°C and maintain while adding the Hydroxypropyl Methylcellulose, slowly, and mix until smooth. Maintain 50°C and add Part B, slowly. while mixing until smooth and uniform. Maintain 50°C, melt Part C and add to Parts A+B with moderate mixing. Remove heat and add the first two Part D ingredients in the order shown, mixing until smooth and uniform. Add the remaining Part D ingredients and mix until uniform and cool.

ANTIDANDRUFF SHAMPOO

Supplier: Sutton Laboratories

	Ingredient:	Wt.%
	Water	49.95
A	Hydroxypropyl Guar	0.50
	Magnesium Aluminum Silicate	0.50
B	Ammonium Lauryl Sulfate	40.00
	PPG-5-Ceteth-10 Phosphate	1.80
	Lauramide DEA	4.00
	Zinc Pyrithione, 48%	2.10
C	Propylene Glycol(and)Diazolidinyl Urea(and) Methylparaben(and)Propylparaben (**Germaben® II**)	1.00
	Fragrance	0.15
	Color	q.s.
		100.00

Mixing Procedure:

Add the Magnesium Aluminum Silicate to the Water slowly with good shear until smooth. Add the remaining Part A ingredient and mix until uniformly dispersed. With slow agitation add Part B. Add The first Part C ingredient and mix until dissolved. Add the remaining Part C ingredients in the order shown. No heating is required.

ANTI-DANDRUFF SHAMPOO

Supplier: R.T. Vanderbilt Company, Inc.

	Ingredient:	Wt.%
A	Tromethamine Magnesium Aluminum Silicate (**Veegum PRO**)	1.20
	Tetrasodium Pyrophosphate (TSPP)	0.06
	Deionized Water	51.64
B	Sodium Laureth Sulfate	40.00
	Myristamide DEA	5.00
C	Zinc Pyrithione, 48%	2.10
	Citric Acid to pH 7	q.s.
	Preservative, Color, Fragrance	q.s.
		100.00

Mixing Procedure:

Add TSPP to the Water and stir until dissolved. Add **Veegum PRO** and mix with maximum available shear until smooth. Heat to 70°C. Melt Myristamide DEA. With slow stirring, add the Part B ingredients in the order shown, maintaining 70°C. Mix until smooth and uniform. Cool to 30°C and, with slow stirring, add the Part C ingredients in the order shown.

ANTIDANDRUFF SULFUR SHAMPOO

Supplier: R.T. Vanderbilt Company, Inc.

	Ingredient:	Wt.%
A	Magnesium Aluminum Silicate (**Veegum**)	1.00
	Water	53.30
B	Colloidal Sulfur	1.00
C	Ammonium Lauryl Sulfate	40.00
	Lauramide DEA	4.50
	Sodium Chloride	0.20
	Preservative	q.s.
		100.00

Mixing Procedure:

Slowly add **Veegum** to the Water, while agitating at maximum available shear until smooth and uniform. Add Part B and Part C ingredients in order shown, mixing each until smooth and uniform.

FREQUENT-USE, ANTIDANDRUFF SHAMPOO

Supplier: R.T. Vanderbilt Company, Inc.

	Ingredient:	Wt.%
A	Purified Bentonite (**Veegum HS**)	1.00
	Deionized Water	47.35
	Citric Acid	0.05
	Hydroxypropyl Guar	1.00
B	Deionized Water	15.00
	Cocoyl Sarcosine (**Vanseal CS**)	3.00
	Ammonium Lauryl Sulfate	30.00
	Polyquaternium-10	0.50
C	Sodium Hydroxide, 25% to pH 6.5	q.s.
	Zinc Pyrithione, 48%	2.10
	Preservative, Color, Fragrance	q.s.
		100.00

Mixing Procedure:

Heat Part A Water to 50°C, add Veegum HS and stir with a propeller mixer at 1800 rpm for 60 minutes. Cool and add the remaining Part A ingredients in the order shown and mix for 15 minutes. In a separate vessel, add Polyquaternium-10 to the Part B Water and mix until a clear solution is obtained. Add the remaining Part B ingredients with slow mixing until clear and uniform. Slow the mixer to 500 rpm. and add Part B mixture to Part A mixture. Continue mixing and cooling and add Part C ingredients in the order shown, mixing each until uniform.

SELENIUM SULFIDE SHAMPOO

Supplier: R.T Vanderbilt Company, Inc.

	Ingredient:	Wt.%
	Purified Bentonite (**Veegum HS**)	1.20
	Water	30.00
A	Titanium Dioxide	1.00
	Propylene Glycol	1.00
	Citric Acid	0.08
	Captan (**Vancide 89RE**)	0.25
	Hydroxypropylcellulose	1.20
B	Selenium Sulfide	1.00
	Water	40.77
	Sodium Lauroyl Sarcosinate (**Vanseal NALS-30**)	8.50
C	Cocamidopropyl Betaine	15.00
	Color, Fragrance	q.s.
		100.00

Mixing Procedure:

Slowly add **Veegum HS** to the Part A Water, agitating with maximum available shear until smooth and uniform. Add the remaining Part A ingredients in the order shown, mixing after each addition until smooth and uniform. Prepare Part B by grinding the Hydroxypropylcellullose/Selenium Sulfide dry blend. Add this dry blend to the Part B Water and mix gently until smooth and uniform. Slowly add the Part B mixture to the Part A mixture and mix gently until uniform. Add the Part C ingredients to Parts A+B, in order, mixing until uniform. Final pH should be adjusted to 6.0 or slightly less.

CHAPTER 13

EXTERNAL ANALGESICS

The formulations in this chapter include creams, lotions, gels, sticks and aerosol sprays designed to deliver analgesics i.e. pain relievers to the skin. The products are used to provide relief of pain from back ache, arthritis, bumps, bruises, teething, etc. The brand name formulations usually contain one or more of the following active ingredients: Trolamine Salicylate, Methyl Salicylate, Benzocaine, Camphor, Menthol.

PART I

BRAND NAME FORMULATIONS

ASPERCREAM

Manufacturer: Thompson Medical Company, Inc.

Active Ingredient & Concentration:
10% Trolamine Salicylate

Product Forms:
Cream
Lotion

Other Ingredients:
Cream contains Cetyl Alcohol, Glycerin, Methylparaben, Mineral Oil Potassium Phosphate, Propylparaben, Stearic Acid, Triethanolamine, Water. Lotion contains Cetyl Alcohol, Fragrance, Glyceryl Stearate, Isopropyl Palmitate, Lanolin, Methylparaben, Potassium Phosphate, Propylene Glycol, Propylparaben, Sodium Lauryl Sulfate, Stearic Acid, Water.

AURUM

Manufacturer: Au Pharmaceuticals, Inc.

Active Ingredients & Concentration:
10% Methyl Salicylate
+ 2.5% Camphor + 3% Menthol

Product Form:
Lotion

Other Ingredients:
Deionized Water, C12-15 Alcohols, Propylene Glycol, Eucalyptus Oil, Stearic Acid, DEA Cetyl Phosphate, PEG-8 Distearate, Carboxy Polymethylene, Methyl Hydroxybenzoate, Ginseng, Imidazolidinyl Urea, Triethanolamine, Edetate Disodium, Urea, Jojoba Oil, Vitamins A, D_3 and E, FD&C Yellow No. 5, 24 Karat Gold, Aloe.

AMERICAINE

Manufacturer: Fisons Consumer Health

Active Ingredient & Concentration:
20% Benzocaine

Product Forms:
Liquid Spray
Ointment

Other Ingredients:
Liquid spray contains Butane, Isobutane, Polyethylene Glycol 200, Propane.
Ointment contains Benzethonium Chloride, Polyethylene Glycol 300, Polyethylene Glycol 3350.

ANBESOL

Manufacturer: Whitehall Laboratories, Inc.

Active Ingredients & Concentration:
6.3% Benzocaine + 0.5% Phenol

Product Forms:
Gel
Liquid

Other Ingredients:
Gel contains Alcohol (70%), Camphor, Carbomer 934P, Colors, Flavor, Glycerin.
Liquid contains Alcohol (70%), Camphor, Glycerin, Menthol, Potassium Iodide, Povidone Iodine.

ANBESOL, MAXIMUM STRENGTH

Manufacturer: Whitehall Laboratories, Inc.

Active Ingredient & Concentration:
20% Benzocaine

Product Forms:
Gel
Liquid

Other Ingredients:
Gel contains Alcohol (60%), Carbomer 934P, Colors, Flavor, PEG-12, Saccharin.
Liquid contains Alcohol (60%), Flavor, PEG-8, Saccharin, Colors.

BABY ANBESOL

Manufacturer: Whitehall Laboratories, Inc.

Active Ingredient & Concentration:
7.5% Benzocaine

Product Form:
Gel

Other Ingredients:
Carbomer 934P, Colors, Edetate Disodium, Flavor, Glycerin, PEG-8, Saccharin, Water. May also contain Methylparaben, Propylparaben, Benzoic Acid.

BEN-GAY, ORIGINAL FORMULA

Manufacturer: Pfizer Consumer Health Care

Active Ingredients & Concentration:
18.3% Methyl Salicylate
+ 16% Menthol

Product Form:
Ointment

Other Ingredients:
Lanolin, Microcrystalline Wax, Purified Water, Synthetic Beeswax.

BEN-GAY, REGULAR STRENGTH

Manufacturer: Pfizer Consumer Health Care

Active Ingredients & Concentration:
15% Methyl Salicylate
+10% Menthol

Product Form:
Ointment

Other Ingredients:
Glyceryl Stearate SE, Anhydrous Lanolin, Polysorbate 85, Sorbitan Tristearate, Stearic Acid, Triethanolamine Stearate, Purified Water.

BEN-GAY, DAYTIME

Manufacturer: Pfizer Consumer Health Care

Active Ingredients & Concentration:
2.5% Menthol

Product Form:
Ointment

Other Ingredients:
Camphor, Diazolidinyl Urea, Isopropyl Alcohol, Nonoxynol-9, Potassium Carbomer 940, Purified Water.

BEN-GAY, EXTRA STRENGTH ARTHRITIS RUB

Manufacturer: Pfizer Consumer Health Care

Active Ingredients & Concentration:
30% Methyl Salicylate
+ 8% Menthol

Product Form:
Ointment

Other Ingredients:
Glyceryl Stearate SE, Anhydrous Lanolin, Polysorbate 85, Sorbitan Tristearate, Stearic Acid, Potassium Stearate, Stearic Acid, Purified Water, Xanthan Gum.

BICOZENE

Manufacturer: Sandoz Pharmaceuticals/Consumer Div.

Active Ingredients & Concentration:
6% Benzocaine
+ 1.67% Resorcinol

Product Form:
Cream

Other Ingredients:
Castor Oil, Chlorthymol, Ethanolamine Stearates, Glycerin, Glyceryl Borate, Glyceryl Stearates, Parachlorometaxylenol, Polysorbate 80, Sodium Stearate, Triglycerol Diisostearate, Fragrance.

CAMPHO-PHENIQUE

Manufacturer: Sterling Winthrop, Inc.

Active Ingredients & Concentration:
10.8% Camphor
+ 4.7% Phenol

Product Forms:
Gel
Liquid

Other Ingredients:
Gel contains Colloidal Silicon Dioxide, Eucalyptus Oil, Glycerin, Mineral Oil.
Liquid contains Eucalyptus Oil, Mineral Oil.

EUCALYPTAMINT, ARTHRITIS PAIN

Manufacturer: CIBA Consumer Pharmaceuticals

Active Ingredient & Concentration:
15% Menthol

Product Form:
Ointment

Other Ingredients:
Lanolin, Eucalyptus Oil.

EUCALYPTAMINT, MUSCLE PAIN

Manufacturer: CIBA Consumer Pharmaceuticals

Active Ingredient & Concentration:
8% Menthol

Product Form:
Liquid

Other Ingredients:
Carbomer 980, Eucalyptus Oil. Fragrance, Propylene Glycol, Alcohol SD3A, Triethanolamine, Polysorbate 80, Water.

FEMININE GOLD

Manufacturer: Au Pharmaceuticals, Inc.

Active Ingredients & Concentration: 3% Menthol + 2.5% Camphor

Product Form: Lotion

Other Ingredients:
Deionized Water, C12-15 Alcohols, Propylene Glycol, Eucalyptus Oil, Stearic Acid, DEA Cetyl Phosphate, PEG-8 Distearate, Carboxy Polymethylene, Methyl Hydroxybenzoate, Ginseng, Imidazolidinyl Urea, Triethanolamine, Edetate Disodium, Urea, Jojoba Oil, Vitamins A, D_3 and E, FD&C Yellow No. 5, 24 Karat Gold, Aloe.

FLEX-ALL 454

Manufacturer: Chattem Consumer Products

Active Ingredient & Concentration: 7% Menthol

Product Form: Lotion

Other Ingredients:
Alcohol, Allantoin, Aloe Vera Gel, Boric Acid, Carbomer 940, Diazolidinyl Urea, Eucalyptus Oil, Glycerin, Iodine, Methylparaben, Methyl Salicylate, Peppermint Oil, Polysorbate 60, Potassium Iodide, Propylene Glycol, Propylparaben, Thyme Oil, Triethanolamine, Water.

FLEX-ALL 454, MAXIMUM STRENGTH

Manufacturer: Chattem Consumer Products

Active Ingredient & Concentration: 16% Menthol

Product Form: Lotion

Other Ingredients:
Allantoin, Aloe Vera Gel, Boric Acid, Carbomer 940, Diisopropyl Adipate, Eucalyptus Oil, Glycerin, Methyl Salicylate, Peppermint Oil, Alcohol SD38B, Steareth-2, Steareth-21, Thyme Oil, Triethanolamine, Water.

ICY HOT

Manufacturer: Chattem Consumer Products

Active Ingredients & Concentration:
Ointment:29% Methyl Salicylate + 7.6% Menthol
Cream and Stick: 30% Methyl Salicylate + 10% Menthol

Product Forms:
Ointment
Cream
Stick

Other Ingredients:
Ointment contains Paraffin, White Petrolatum.
Cream contains Carbomer, Cetyl Esters Wax, Emulsifying Wax, Oleth-3 Phosphate, Stearic Acid, Trolamine, Water.
Stick contains Ceresin, Cyclomethicone, Hydrogenated Castor Oil, Microcrystalline Wax, Paraffin, PEG-150 Distearate, Propylene Glycol, Stearic Acid, Stearyl Alcohol.

MYOFLEX

Manufacturer: Fisons Consumer Health

Active Ingredient & Concentration:
10% Trolamine Salicylate

Product Form:
Cream

Other Ingredients:
Cetyl Alcohol, Edetate Disodium, Fragrance, Propylene Glycol, Purified Water, Sodium Lauryl Sulfate, Stearyl Alcohol, White Wax.

MOBISYL

Manufacturer: B.F. Ascher & Company, Inc.

Active Ingredient & Concentration:
10% Trolamine Salicylate

Product Form:
Cream

Other Ingredients:
Glycerin, Methylparaben, Mineral Oil, Polysorbate 60, Propylparaben, Sorbitan Stearate, Sorbitol, Stearic Acid, Water.

NUPERCAINAL, PAIN RELIEF

Manufacturer: CIBA Consumer Pharmaceuticals

Active Ingredient & Concentration:	Product Form:
0.5% Dibucaine	Cream

Other Ingredients:
Acetone, Sodium Bisulfite, Fragrance, Glycerin, Potassium Hydroxide, Purified Water, Stearic Acid, Trolamine.

RHULICREAM

Manufacturer: Rydelle Laboratories

Active Ingredients & Concentration:	Product Form:
3.0% Calamine + 0.3% Camphor + 1.0% Pramoxine Hydrochloride	Cream

Other Ingredients:
Glycerin, Distearyldimonium Chloride, Petrolatum, Isopropyl Palmitate, Cetyl Alcohol, Dimethicone, Sodium Chloride, Water.

RHULIGEL

Manufacturer: Rydelle Laboratories

Active Ingredients & Concentration:	Product Form:
0.3% Menthol + 0.3% Camphor + 2% Benzyl Alcohol	Gel

Other Ingredients:
Alcohol SD23A (31% w/w), Purified Water, Propylene Glycol, Carbomer 940, Triethanolamine, Benzophenone-4, EDTA.

RHULISPRAY

Manufacturer: Rydelle Laboratories

Active Ingredients & Concentration:
13.8% Calamine + 0.7% Camphor
+ 5.0% Benzocaine

Product Form:
Liquid Spray

Other Ingredients:
Benzyl Alcohol, Hydrated Silica, Isobutane, Isopropyl Alcohol (70% w/w), Oleyl Alcohol, Sorbitan Trioleate.

SPORTSCREME

Manufacturer: Thompson Medical Company, Inc.

Active Ingredient & Concentration:
10% Trolamine Salicylate

Product Form:
Cream

Other Ingredients:
Cetyl Alcohol, FD&C Blue No. 1, FD&C Yellow No. 5, Fragrance, Glycerin, Methylparaben, Mineral Oil, Potassium Phosphate Monobasic, Propylparaben, Stearic Acid, Triethanolamine, Water,

---

THERAGOLD

Manufacturer: Au Pharmaceuticals, Inc.

Active Ingredients & Concentration:
10% Methyl Salicylate
+ 2.5% Camphor + 3% Menthol

Product Form:
Lotion

Other Ingredients:
Deionized Water, C12-15 Alcohols, Propylene Glycol, Eucalyptus Oil, Stearic Acid, DEA Cetyl Phosphate, PEG-8 Distearate, Carboxy Polymethylene, Methylparaben, Ginseng, Imidazolidinyl Urea, Triethanolamine, Edetate Disodium, Urea, Jojoba Oil, Vitamins A & D_3 and E, FD&C Yellow No. 5, 24 Karat Gold, Aloe.

THERAPEUTIC GOLD

Manufacturer: Au Pharmaceuticals, Inc.

Active Ingredients & Concentration:
10% Methyl Salicylate + 3% Menthol + 2.5% Camphor

Product Form:
Lotion

Other Ingredients:
Deionized Water, C12-15 Alcohols, Propylene Glycol, Eucalyptus Oil, Stearic Acid, DEA Cetyl Phosphate, PEG-8 Distearate, Carboxy Polymethylene, Methyl Hydroxybenzoate, Ginseng, Imidazolidinyl Urea, Triethanolamine, Edetate Disodium, Urea, Jojoba Oil, Vitamins A & D_3 and E, FD&C Yellow No. 5, 24 Karat Gold, Aloe.

THERAPEUTIC MINERAL ICE

Manufacturer: Bristol-Myers Products

Active Ingredients & Concentration:
Regular Strength: 2% Menthol
Exercise Formula: 4% Menthol

Product Form:
Gel

Other Ingredients:
Both products contain Ammonium Hydroxide, Carbomer 934P, Cupric Sulfate, FD&C Blue No. 1, Isopropyl Alcohol, Magnesium Sulfate, Sodium Hydroxide, Thymol, Water. May also contain Fragrance, Carbomer 934.

XYLOCAINE

Manufacturer: Astra Pharmaceutical Products, Inc.

Active Ingredient & Concentration:
2.5% Lidocaine

Product Form:
Ointment

Other Ingredients:
Polyethylene Glycols, Propylene Glycol.

PART II

SUPPLIER'S FORMULATIONS

FIRST AID SUNBURN SPRAY

Supplier: Amerchol Corp.

	Ingredient:	Wt.%
A	Allantoin	0.10
B	Deionized Water	35.00
C	Benzocaine	2.00
D	Alcohol SD40	47.90
E	Methyl Gluceth-10 (**Glucam E-10**)	10.00
F	Aloe Vera	5.00
		100.00

Mixing Procedure:

Dissolve Part A in Part B at 35°C. In a separate vessel, dissolve Part C in Part D with agitation. Add Part E, then Parts A + B, mixing well after each addition. When the solution is clear, add Part F. Mix until homogeneous. Chill for 24 hours and filter before packaging.

ANALGESIC CREAM

Supplier: American Colloid Company

	Ingredient:	Wt.%
	Hectorite (**Hectabrite DP**)	1.75
A	Hydroxypropylcellulose	1.75
	Water	40.75
	Ethanol	36.75
	Eucalyptus Oil	0.20
	Menthol	0.40
	Peppermint Oil	0.40
B	Triethanolamine	3.00
	Methyl Salicylate	5.00
	Cocoyl Sarcosine	10.00
	Preservative	q.s.
		100.00

Mixing Procedure:

Dry blend **Hectabrite DP** and Hydroxypropylcellulose and add to the Water and Ethanol slowly, agitating continually until smooth. Combine the Part B ingredients and mix until all are dissolved. Add Part B to Part A with slow mixing. Mix until uniform.

ANALGESIC GEL

Supplier: American Distilling & Manufacturing, Inc.

	Ingredient:	Wt.%
A	Sodium Magnesium Silicate	2.50
B	Water	33.75
	Alcohol SD39C	25.00
	Isopropyl Myristate	5.00
C	Menthol	6.00
	Camphor	3.25
	Distilled Witch Hazel Extract, 14% Alcohol	24.50
		100.00

Mixing Procedure:

Disperse Part A in Part B with high shear. Add the Part C ingredients in the order shown and mix each until uniform.

CLEAR ANALGESIC STICK

Supplier: Croda, Inc.

	Ingredient:	Wt.%
	Methyl Salicylate	35.00
A	Menthol	15.00
	PPG-3 Myristyl Ether (**Promyristyl PM-3**)	32.00
	Cocamide DEA Cocoyl Sarcosine (**Incromide CAC**)	7.00
B	Sodium Stearate	8.00
C	Deionized Water	3.00
		100.00

Mixing Procedure:

Combine the Part A Ingredients with mixing and heat to 90°C. Dust Part B into Part A with mixing. When uniform, add Part C to the batch and fill above set point (~85°C).

HEAT-RUB LOTION

Supplier: Croda, Inc.

	Ingredient:	Wt.%
	PPG-5-Ceteth-10 Phosphate (**Crodafos SG**)	2.50
	Emulsifying Wax NF (**Polawax**)	4.00
A	PPG-10 Cetyl Ether (**Procetyl 10**)	4.50
	Menthol	1.00
	Camphor	1.00
	Methyl Salicylate	7.50
	Glycerin	3.00
B	Gelatin NF (**Crodyne BY-19**)	1.00
	Diethanolamine	0.30
	Deionized Water	74.20
C	Propylene Glycol(and)Diazolidinyl Urea (and)Methylparaben(and)Propylparaben	1.00
		100.00

Mixing Procedure:

Premix Menthol, Camphor and Methyl Salicylate with **Procetyl 10**. When completely dissolved, add the remaining Part A ingredients and heat to 75-80°C. Dissolve **Crodyne BY-19** in the water and add the remaining Part B ingredients. Heat to 80°C. Add Part B to Part A slowly, using good mechanical agitation. Allow to cool, while mixing, to 40°C and add Part C. Continue mixing, cool to 30°C and fill.

MEDICATED, GREASELESS SKIN CREAM

Supplier: Croda, Inc.

	Ingredient:	Wt.%
A	Stearic Acid, XXX*	16.00
	Lanolin USP (**Super Corona Lanolin**)	4.00
B	Triethanolamine, 99%	0.50
	Gelatin (**Crodyne BY-19**)	2.00
	Lime Water**	30.00
	Deionized Water	47.20
C	"Medication Mix"***	0.30
		100.00

* XXX is a common abbreviation for triple-pressed Stearic Acid

** Saturated Calcium Hydroxide solution either filtered or decanted.

*** Composition: 15.0% Menthol, 30.0% Camphor, 10.0% Oil of Clove, 15.0% Oil of Eucalyptus, 30.0% Phenol.

Mixing Procedure:

Combine the Part A ingredients with mixing and heat to 60-65°C. Combine the Part B ingredients with mixing and heat to 60-65°C. Add Part B to Part A with good mixing. Cool to 45°C with mixing and add Part C. Cool to the desired fill temperature.

INSECT BITE TREATMENT CREAM

Supplier: Gattefossé s.a.

	Ingredient:	Wt.%
	Trilaneth-4 Phosphate(and)Glyceryl Stearate (and)PEG-2 Stearate (**Sedefos 75**)	18.00
	Hydrogenated Palm/Palm Kernel Oil PEG-6 Esters (**Labrafil M 2130 CS**)	2.00
A	Mineral oil	8.00
	Sodium Methylparaben	0.03
	Sorbic Acid	0.07
	Demineralized Water	64.67
	Benzocaine	1.00
	Butamben	1.00
B	Menthol	0.20
	Resorcinol	0.03
	Ethoxydiglycol (**Transcutol**)	5.00
		100.00

Mixing Procedure:

Dissolve the first four Part B ingredients in the **Transcutol**. Mix and heat Part A to 75°C. Allow to cool slowly with constant stirring. At 35°C, add the Part B mixture to Part A. Homogenize, if necessary.

FOOT CREAM

Supplier: International Sourcing, Inc.

Ingredient:	Wt.%
Octyldodecanol	10.00
Mineral Oil	6.00
Camphor	0.50
Deionized Water	78.35
Coltsfoot Extract (**Phytelene EG-512**)	2.00
Pine Tree Extract (**Phytelene EG-028**)	2.00
Walnut Extract (**Phytelene EG-462**)	1.00
Fragrance	0.15
Preservative	q.s.
	100.00

Mixing Procedure:

Mix all of the above ingredients in the water, heated to 50°C, until uniform

---

EUCALYPTUS + MINT EMULSION, SUPER RICH

Supplier: R.I.T.A. Corp.

	Ingredient:	Wt.%
A	Distilled Water	42.25
	Eucalyptamint	37.50
	Sodium Stearoyl Lactylate (**Pationic SSL**)	7.00
B	PEG-20 Hydrogenated Lanolin (**Supersat AWS 4**)	3.50
	Ritasynt IP	1.75
	Cetearyl Alcohol(and)Polysorbate 60(and) PEG-15 Stearate(and)Steareth-20 (**Ritachol 1000**)	8.00
		100.00

Mixing Procedure:

Heat Part A to 71°C. Combine the Part B ingredients in another container and heat to 71°C. Slowly add Part A to Part B and mix for 1 hour. Cool the mixture to 35-45°C. Pour into containers.

EUCALYPTUS + MINT TYPE OINTMENT

Supplier: R.I.T.A. Corp.

Ingredient:	Wt.%
Menthol	16.00
Eucalyptus	4.00
Lanolin Anhydrous, USP	80.00
	100.00

Mixing Procedure:

Heat the lanolin until melted (approximately 50°C). Add the remaining ingredients with mixing. Mix for a minimum of 1 hour. Fill hot.

10% T.E.A. SALICYLATE CREAM

Supplier: Sutton Laboratories

	Ingredient:	Wt.%
	Glyceryl Stearate	5.00
	Cetyl Alcohol	2.50
A	Cetyl Phosphate(and)DEA Cetyl Phosphate	3.00
	Stearyl Stearoyl Stearate	4.00
	Coco-Caprylate/Caprate	4.00
	Cetyl Palmitate	4.00
	Dimethicone	0.50
	Deionized Water	50.20
B	Propylene Glycol(and) Diazolidinyl Urea((and) Methylparaben(and)Propylparaben (**Germaben® II**)	1.00
C	Magnesium Aluminum Silicate	0.55
	Xanthan Gum	0.25
	Deionized Water	10.00
D	Trolamine Salicylate (TEA-Salicylate)	10.00
	Propylene Glycol	5.00
		100.00

Mixing Procedure:

Add the Part C ingredients to the Part B ingredients (heated to 85°C). Mix until smooth and all are dispersed. Then add the Part A ingredients to Parts B+C and mix well at 80-85°C. Continue mixing, while cooling, to 60°C. Add the Part D ingredients, mix well and continue mixing and cooling to 35°C. Desired pH is 5.5-5.6.

ANALGESIC CREAM

Supplier: R.T. Vanderbilt Company, Inc.

	Ingredient:	Wt.%
A	Tromethamine Magnesium Aluminum Silicate (**Veegum PRO**)	3.00
	Hydroxypropylcellulose	3.00
	Water	34.90
	Ethanol	34.90
B	Cocoyl Sarcosine (**Vanseal CS**)	4.00
	Oleth-10	2.50
	PEG-25 Hydrogenated Castor Oil	2.50
	Isopropyl Myristate	5.00
	Triethanolamine	2.00
C	Camphor	0.50
	Menthol	0.50
	Peppermint Oil	0.50
	Eucalyptus Oil	0.20
	Methyl Salicylate	6.50
	Preservative	q.s.
		100.00

Mixing Procedure:

Dry blend **Veegum PRO** and Hydroxypropylcellulose and add them to the water and ethanol slowly, agitating continually until smooth. Combine the Part B ingredients. Combine the Part C ingredients and add to Part B with slow stirring. Add Parts B+C to Part A with slow mixing. Mix until uniform.

BURN/BITE CREAM

Supplier: R.T. Vanderbilt Company, Inc.

	Ingredient:	Wt.%
	Magnesium Aluminum Silicate (**Veegum**)	1.50
A	Water	56.80
	Propylene Glycol	3.00
	Dimethicone Emulsion	0.20
	Mineral Oil, Light	10.00
	Acetylated Lanolin Alcohol	17.00
	Menthol, USP	1.00
B	Benzocaine, USP	5.00
	C18-36 Acid	3.00
	Glyceryl Stearate(and)PEG-100 Stearate	12.00
	Polysorbate 60	0.50
	Preservatives	q.s.
		100.00

Mixing Procedure:

Add the **Veegum** to the water slowly, agitating with maximum available shear until smooth. Add the remaining Part A ingredients and heat to 75-80°C. Mix and heat the Part B ingredients, keeping Benzocaine suspended, to 75-80°C. Add Part B to Part A and mix while cooling. Pour off at about 40°C.

ANTI-INFLAMMATORY CREAM

Supplier: R.T. Vanderbilt Company, Inc.

	Ingredient:	Wt.%
	Palmitoyl Collagen Amino Acids	2.00
	Isopropyl Palmitate	10.00
A	Tristearin	5.00
	Squalane	5.00
	Glyceryl Stearate SE	10.00
	Deionized Water	63.00
B	Glycerin	5.00
	Preservative, Dye, Fragrance	q.s.
		100.00

Mixing Procedure:

Blend Part A ingredients and heat to 70°C. Mix until uniform. Mix the Part B ingredients and heat to 70°C. Slowly add Part A to Part B with high shear mixing. Avoid incorporating air. Cool with continuous mixing to 30°C, add Part C and mix until uniform.

EXTERNAL ANALGESIC CREAM

Supplier: R.T. Vanderbilt Company, Inc.

	Ingredient:	Wt.%
A	Magnesium Aluminum Silicate (**Veegum**)	1.50
	Water	54.70
	Simethicone Emulsion	0.20
	Propylene Glycol	3.00
B	Methyl Salicylate	15.00
	Menthol	5.00
	Polysorbate 80	0.60
	C18-36 Acid	5.00
	Glyceryl Stearate(and)PEG-100 Stearate	15.00
C	Preservative	q.s.
		100.00

Mixing Procedure:

Add the **VEEGUM** to the Water slowly, agitating with maximum available shear until smooth. Add the additional Part A ingredients in the order shown, mixing each until uniformly dispersed. Heat Part A to 75-80°C. Mix and heat the Part B ingredients to 75-80°C. Add Part B to Part A and mix while cooling. Add Part C, mix and pour off at 40°C.

EXTERNAL ANALGESIC LOTION

Supplier: R.T. Vanderbilt Company, Inc.

	Ingredient:	Wt.%
	Magnesium Aluminum Silicate (**Veegum**)	0.75
	Water	72.85
A	Simethicone Emulsion	0.10
	Nonfat Dry Milk(and)Xanthan Gum(and) Glyceryl Stearate(and)Propylene Glycol Alginate(and)Hydrogenated Vegetable Glycerides Phosphate	0.70
	Propylene Glycol	3.00
	Methyl Salicylate	15.00
	Menthol	5.00
B	Polysorbate 80	0.60
	C12-15 Alcohols Benzoate	2.00
C	Preservative	q.s.
		100.00

Mixing Procedure:

Add the **VEEGUM** to the Water, slowly, agitating with maximum available shear until smooth. Add the additional Part A ingredients in the order shown, mixing each until uniformly dispersed. Mix Part B ingredients until clear. Add the Part B ingredients to Part A and mix until uniform. Add Part C, mix and pour off.

CHAPTER 14

HEMORRHOIDAL TREATMENTS

PART I

BRAND NAME FORMULATIONS

AMERICAINE

Manufacturer: Fisons Consumer Health

Active Ingredient & Concentration
20% Benzocaine

Product Form:
Ointment

Other Ingredients:
Benzethonium Chloride, Polyethylene Glycol 300, Polyethylene Glycol 3350.

ANUSOL

Manufacturer: Parke-Davis

Active Ingredients & Concentration
Suppository: 88.7% Hard Fat + 0.25% Phenylephrine Hydrochloride
Ointment: 1% Pramoxine Hydrochloride + 12.5% Zinc Oxide

Product Forms:
Suppository
Ointment

Other Ingredients:
Suppositories contain Corn Starch, Methylparaben, Propylparaben.
Ointment contains Mineral Oil (46.7%), Benzyl Benzoate, Calcium Phosphate Dibasic, Cocoa Butter, Glyceryl Monooleate, Glyceryl Monostearate, Kaolin, Peruvian Balsam, Polyethylene Wax.

ANUSOL HC-1

Manufacturer: Parke-Davis

Active Ingredient & Concentration:
Hydrocortisone Acetate equal to 1% Hydrocortisone

Product Form:
Ointment

Other Ingredients:
Diazolidinyl Urea, Methylparaben, Microcrystalline Wax, Mineral Oil, Propylene Glycol, Propylparaben, Sorbitan Sesquioleate, White Petrolatum.

NUPERCAINAL

Manufacturer: CIBA Consumer Pharmaceuticals

Active Ingredients & Concentration
Ointment: 1% Dibucaine
Suppository: 2.1gm Cocoa Butter + 0.25gm Zinc Oxide/Suppository
Cream: 0.5% Dibucaine

Product Forms:
Ointment
Suppository
Cream

Other Ingredients:
Ointment contains Acetone, Sodium Bisulfite, Lanolin, Light Mineral Oil, Purified Water, White Petrolatum.
Suppository contains Acetone, Sodium Bisulfite, Bismuth Subgallate.
Cream contains Acetone, Sodium Bisulfite, Fragrance, Glycerin, Potassium Hydroxide, Purified Water, Stearic Acid, Trolamine.

PAZO

Manufacturer: Bristol-Myers Products

Active Ingredients & Concentration
Ointment: 2% Camphor + 0.2% Ephedrine Sulfate + 5% Zinc Oxide
Suppository: 3.86mg Ephedrine Sulfate + 96.5mg Zinc Oxide/Suppository

Product Forms:
Ointment
Suppository

Other Ingredients:
Ointment contains Lanolin, Petrolatum.
Suppository contains Hydrogenated Vegetable Oil.

PREPARATION H

Manufacturer: Whitehall Laboratories, Inc.

Active Ingredients & Concentration
- Ointment or Cream: 2000 Units of Skin Respiratory Factor/Ounce + 3% Shark Liver Oil
- Suppository: 2000 Units of Skin Respiratory Factor/Ounce + 3% Shark Liver Oil

Product Forms:
- Ointment
- Cream
- Suppository

Other Ingredients:
- Ointment contains Beeswax, Glycerin, Lanolin, Lanolin Alcohol, Mineral Oil, Paraffin, Phenylmercuric Nitrate, Thymol.
- Cream contains BHA, Cellulose Gum, Cetyl Alcohol, Citric Acid, Edetate Disodium, Glycerin, Glyceryl Stearate, Lanolin, Methylparaben, Phenylmercuric Nitrate, Propyl Gallate, Propylene Glycol, Propylparaben, Simethicone, Sodium Lauryl Sulfate, Stearyl Alcohol, Water, Xanthan Gum. May also contain Glyceryl Oleate, Polysorbate 80.
- Suppository contains Beeswax, Glycerin, Phenylmercuric Nitrate, PEG-600-Dilaurate.

TRONOLANE

Manufacturer: Ross Laboratories

Active Ingredients & Concentration
- Cream: 1% Pramoxine Hydrochloride
- Suppository: 5% Zinc Oxide + 95% Hard Fat

Product Forms:
- Cream
- Suppository

Other Ingredients:
- Cream contains Beeswax, Cetyl Alcohol, Cetyl Esters Wax, Glycerin, Methylparaben, Propylparaben, Sodium Lauryl Sulfate, Zinc Oxide.
- Suppository contains none.

TUCKS

Manufacturer: Parke-Davis

Active Ingredient & Concentration
Cream or Pad: 50% Witch Hazel

Product Forms:
Cream
Pad

Other Ingredients:
Cream contains Alcohol (7%), White Petrolatum, Sorbitol, Cetyl Alcohol, Polysorbate 60, Polyethylene Stearate, Anhydrous Lanolin, Glyceryl Oleate, Propylene Glycol, Benzethonium Chloride.
Pad contains Alcohol (7%), Sodium Citrate, Citric Acid, Methylparaben, Water, Glycerin, Aloe Vera Gel, Propylparaben, Propylene Glycol, Diazolidinyl Urea.

WYANOIDS

Manufacturer: Wyeth-Ayerst Laboratories

Active Ingredients & Concentration
2000 Units of Skin Respiratory Factor/Ounce + 3% Shark Liver Oil

Product Form:
Suppository

Other Ingredients:
Beeswax, Cocoa Butter, Glycerin, Phenylmercuric Nitrate, PEG-600-Dilaurate.

PART II

SUPPLIER'S FORMULATIONS

HEMORRHOIDAL CREAM/OINTMENT

Supplier: Brooks Industries, Inc.

	Ingredient:	Wt.%
A	Lanolin Alcohol (**Ivarlan 3310**)	2.00
	Petrolatum	44.80
	Petrolatum, Amber	45.00
	Shark Liver Oil	3.00
B	Live Yeast Cell Derivative (**Biodynes TRF**)	1.00
	Demineralized Water	1.00
	Lanolin (**Ivarlan 3001**)	3.09
C	Thyme Oil	0.10
	Phenyl Mercuric Nitrate	0.01
		100.00

Mixing Procedure:

In the main vessel, mix and heat the Part A ingredients to 70°C. Cool to 50°C and hold. Combine the Part B ingredients in another vessel, heat to 40°C and mix until uniformly dispersed. With rapid agitation, add Part B to Part A. Mix until uniformly dispersed. Cool to 40°C and add the Part C ingredients. Continue mixing while cooling to 35°C.

ANTI-HEMORRHOIDAL OINTMENT

Supplier: Gattefossé s.a.

	Ingredient:	Wt.%
	Trilaneth-4 Phosphate(and)Glycol Stearate (and)PEG-2 Stearate (**Sedefos 75**)	18.00
A	Hydrogenated Palm/Palm Kernel PEG-6 Esters (**Labrafil M 2130 CS**)	2.00
	Mineral Oil	16.00
	Sorbic Acid	0.07
	Sodium Methylparaben	0.03
	Demineralized Water	71.80
B	Anti-Hemorrhoid Essential Oil	2.10
		100.00

Mixing Procedure:

Mix and heat the Part A ingredients to 75°C. Cool with constant mixing. At 30°C, add Part B. Continue mixing until cool.

ANTI-HEMORRHOIDAL SUPPOSITORIES

Supplier: Gattefossé s.a.

	Ingredient:	Wt.%
	Benzocaine	0.93
A	Bismuth Iodosubgallate	2.33
	Bismuth Subgallate	2.33
	Zinc Oxide	9.33
	Aesculus Extract	0.57
B	Witch Hazel Extract	0.57
	Demineralized Water	0.57
C	**Suppocire AS2X**	83.37
		100.00

Mixing Procedure:

Melt Part C at 50°C. Allow to cool to 40°C and add the Part A ingredients to Part C with turbine mixing. Dissolve the first two Part B ingredients in the Part B water and add the mixture to Part A+C at about 38°C. Casting is carried out at 34°C.

CHAPTER 15

INTERNAL ANALGESICS, ANTIPYRETICS & MENSTRUAL PRODUCTS

For many years aspirin or aspirin combined with caffeine or buffering agents were the only non-prescription, active ingredients available to relieve pain or reduce fever. Then, following World War II, Acetaminophen was added to this short list of pain relievers and , in 1984, the U.S. FDA permitted Ibuprophen to be supplied to consumers in non-prescription strength.

The menstrual products included in this chapter are for the relief of pain due to cramps and may also contain a diuretic, i.e. Parabrom or an antihistamine, i. e. Pyrilamine Maleate.

PART I

BRAND NAME FORMULATIONS

ADVIL

Manufacturer: Whitehall Laboratories, Inc.

Active Ingredient & Concentration:
200mg Ibuprophen/Tablet or Caplet

Product Forms:
Tablet
Caplet

Other Ingredients:
Acacia, Acetylated Monoglycerides, Beeswax or Carnauba Wax, Calcium Sulfate, Colloidal Silicon Dioxide, Dimethicone, Iron Oxide, Lecithin, Pharmaceutical Glaze, Povidone, Sodium Benzoate, Carboxymethylcellulose Sodium, Starch, Stearic Acid, Sucrose, Titanium Dioxide.

ALLEREST, HEADACHE STRENGTH

Manufacturer: Fisons Consumer Health

Active Ingredients & Concentration:
325mg Acetaminophen
+ 30mg Pseudoephedrine Hydrochloride
+ 2mg Chlorpheniramine Maleate/Tablet

Product Form:
Tablet

Other Ingredients:
Magnesium Stearate, Povidone, Microcrystalline Cellulose, Pregelatinized Starch.

ALLEREST, SINUS PAIN FORMULA

Manufacturer: Fisons Consumer Health

Active Ingredients & Concentration:
500mg Acetaminophen
+ 30mg Pseudoephedrine Hydrochloride
+ 2mg Chlorpheniramine Maleate/Tablet

Product Form:
Tablet

Other Ingredients:
Magnesium Stearate, Povidone, Microcrystalline Cellulose, Pregelatinized Starch, Sodium Starch Glycolate.

ALLEREST, NO DROWSINESS

Manufacturer: Fisons Consumer Health

Active Ingredients & Concentration:
325mg Acetaminophen
+ 30mg Pseudoephedrine Hydrochloride/
Tablet

Product Form:
Tablet

Other Ingredients:
Magnesium Stearate, Povidone, Microcrystalline Cellulose, Pregelatinized Starch.

ANACIN

Manufacturer: Whitehall Laboratories, Inc.

Active Ingredients & Concentration:
400mg Aspirin + 32mg Caffeine/Tablet or Caplet

Product Forms:
Tablet
Caplet

Other Ingredients:
Tablets contain Hydroxypropyl Methylcellulose, Microcrystalline Cellulose, Polyethylene Glycol, Sodium Lauryl Sulfate, Starch.
Caplets contain Hydroxypropyl Methylcellulose, Iron Oxide, Microcrystalline cellulose, Polyethylene Glycol, Sodium Lauryl Sulfate, Starch.

ANACIN, MAXIMUM STRENGTH

Manufacturer: Whitehall Laboratories, Inc.

Active Ingredients & Concentration:
500mg Aspirin + 32mg Caffeine/Tablet

Product Form:
Tablet

Other Ingredients:
Hydroxypropyl Methylcellulose, Microcrystalline Cellulose, Polyethylene Glycol, Sodium Lauryl Sulfate, Starch.

ANACIN, ASPIRIN FREE, P.M.

Manufacturer: Whitehall Laboratories, Inc.

Active Ingredients & Concentration:
500mg Acetaminophen
+ 25mg Diphenhydramine Hydrochloride/Caplet

Product Form:
Caplet

Other Ingredients:
Calcium Stearate, FD&C Blue No. 1 Lake, Hydroxypropyl Methylcellulose, Magnesium Stearate, Microcrystalline Cellulose, Polyethylene Glycol, Povidone, Starch, Stearic Acid, Titanium Dioxide. May also contain Ammonium Hydroxide, Iron Oxide, Pharmaceutical Glaze, Potassium Hydroxide, Propylene Glycol.

ANACIN, ASPIRIN FREE

Manufacturer: Whitehall Laboratories, Inc.

Active Ingredient & Concentration:
500mg Acetaminophen/Tablet, Caplet or Gelcap

Product Forms:
Tablet
Caplet
Gelcap

Other Ingredients:
Tablets contain Croscarmellose Sodium, FD&C Blue No. 1 Lake, Hydroxypropyl Methylcellulose, Polyethylene Glycol, Povidone, Propylene Glycol, Starch, Stearic Acid, Titanium Dioxide.
Caplets contain Calcium Stearate, Croscarmellose Sodium, D&C Red No.7 Lake, FD&C Blue No. 1 Lake, Hydroxypropyl Methylcellulose, Polyethylene Glycol, Povidone, Propylene Glycol, Starch, Stearic Acid, Titanium Dioxide.
Gelcaps contain Croscarmellose Sodium, D&C Red No. 27 Lake, Dimethicone, EDTA, FD&C Blue No. 1 Lake, Gelatin, Glycerin, Hydroxypropyl Methylcellulose, Iron Oxide, Lecithin, Polyethylene Glycol, Povidone, Propylene Glycol, Starch, Stearic Acid, Titanium Dioxide, Triacetin, Pharmaceutical Glaze.

BAYER, CHILDREN'S

Manufacturer: Sterling Winthrop, Inc.

Active Ingredient & Concentration:
81mg Aspirin/Tablet

Product Form:
Tablet, Ch.

Other Ingredients:
Dextrose, FD&C Yellow No. 6, Flavor, Saccharin Sodium, Starch.

BAYER, REGULAR & MAXIMUM STRENGTH

Manufacturer: Sterling Winthrop, Inc.

Active Ingredients & Concentration:
- Regular Strength: 325mg Aspirin/Tablet or Caplet
- Maximum Strength: 500mg Aspirin/Tablet or Caplet

Product Forms:
- Tablet
- Caplet

Other Ingredients:
Starch, Triacetin, Hydroxypropyl Methylcellulose.

BAYER, 8-HOUR

Manufacturer: Sterling Winthrop, Inc.

Active Ingredient & Concentration:
650mg Aspirin/Caplet

Product Form:
Caplet, S.R.

Other Ingredients:
Guar Gum, Microcrystalline Cellulose, Starch and other ingredients.

BAYER, ENTERIC, REGULAR & ADULT LOW STRENGTH

Manufacturer: Sterling Winthrop, Inc.

Active Ingredients & Concentration:
- Regular Strength: 325mg Aspirin/Caplet
- Low Strength: 81mg Aspirin/Caplet

Product Form:
Caplet

Other Ingredients:
Regular strength contains D&C Yellow No. 10, FD&C Yellow No. 6, Hydroxypropyl Methylcellulose, Methacrylic Acid Copolymer, Starch, Titanium Dioxide, Triacetin.
Low strength contains Croscarmellose Sodium, D&C Yellow No. 10, FD&C Yellow No. 6, Hydroxypropyl Methylcellulose, Iron Oxide, Lactose, Methacrylic Acid, Microcrystalline Cellulose, Polysorbate 80, Sodium Lauryl Sulfate, Starch, Titanium Dioxide, Triacetin.

BAYER SELECT, IBUPROFEN

Manufacturer: Sterling Winthrop, Inc.

Active Ingredient & Concentration:
200mg Ibuprofen/Caplet

Product Form:
Caplet

Other Ingredients:
Calcium Phosphate, Cellulose, Magnesium Stearate, Silicon Dioxide, Sodium Lauryl Sulfate, Sodium Starch Glycolate, Stearic Acid, Titanium Dioxide.

BAYER SELECT, HEADACHE

Manufacturer: Sterling Winthrop, Inc.

Active Ingredients & Concentration:
500mg Acetaminophen
+ 65mg Caffeine/Caplet

Product Form:
Caplet

Other Ingredients:
Croscarmellose Sodium, FD&C Blue No. 2, Hydroxypropyl Methylcellulose, Magnesium Stearate, Microcrystalline Cellulose, Polyethylene Glycol, Potassium Sorbate, Starch, Titanium Dioxide, Xanthan Gum.

BAYER SELECT, MENSTRUAL

Manufacturer: Sterling Winthrop, Inc.

Active Ingredients & Concentration:
500mg Acetaminophen
+ 25mg Pamabrom/Caplet

Product Form:
Caplet

Other Ingredients:
Croscarmellose Sodium, D&C Red No. 27, Hydroxypropyl Methylcellulose, Iron Oxide, Magnesium Stearate, Microcrystalline Cellulose, Polyethylene Glycol, Polysorbate 80, Starch, Titanium Dioxide.

BAYER SELECT, SINUS PAIN RELIEF

Manufacturer: Sterling Winthrop, Inc.

Active Ingredients & Concentration:	Product Form:
500mg Acetaminophen + 30mg Pseudoephedrine Hydrochloride/ Caplet	Caplet

Other Ingredients:
Colloidal Silicon Dioxide, Colors, Hydroxypropyl Methylcellulose, Iron Oxide, Magnesium Stearate, Microcrystalline Cellulose, Polyethylene Glycol, Polysorbate 80, Povidone, Starch, Titanium Dioxide.

CAMA

Manufacturer: Sandoz Pharmaceuticals/Consumer Div.

Active Ingredients & Concentration:	Product Form:
500mg Aspirin + 150mg Magnesium Oxide + Aluminum Hydroxide Dried Gel equal to 125mg Aluminum Hydroxide/Tablet	Tablet

Other Ingredients:
Colloidal Silicon Dioxide, Croscarmellose Sodium, Hydrogenated Vegetable Oil, Methylcellulose, Methylparaben, Microcrystalline Cellulose, Polyethylene Glycol, Povidone, Pregelatinized Starch, Starch, Colors.

DOAN'S, REGULAR STRENGTH

Manufacturer: CIBA Consumer Pharmaceuticals

Active Ingredient & Concentration:	Product Form:
325mg Magnesium Salicylate/Caplet	Caplet

Other Ingredients:
Magnesium Stearate, Microcrystalline Cellulose, Color, Polyethylene Glycol, Purified Water, Stearic Acid.

DOAN'S, EXTRA STRENGTH

Manufacturer: CIBA Consumer Pharmaceuticals

Active Ingredient & Concentration:
500mg Magnesium Salicylate/Caplet

Product Form:
Caplet

Other Ingredients:
Hydroxypropyl Methylcellulose, Magnesium Stearate, Microcrystalline Cellulose, Polyethylene Glycol, Polysorbate 80, Propylene Glycol, Stearic Acid, Titanium Dioxide.

ECOTRIN, REGULAR & MAXIMUM STRENGTH

Manufacturer: SmithKline Beecham, Inc.

Active Ingredients & Concentration:
Regular Strength: 325mg Aspirin/Tablet or Caplet
Maximum Strength: 500mg Aspirin/Tablet or Caplet

Product Forms:
Tablet
Caplet

Other Ingredients:
Cellulose, Cellulose Acetate Phthalate, D&C Yellow No. 10, Diethyl Phthalate, FD&C Yellow No. 6, Pregelatinized Starch, Silicon Dioxide, Sodium Starch Glycolate, Stearic Acid, Titanium Dioxide and trace amounts of other ingredients.

EMPRIN

Manufacturer: Burroughs Wellcome Company

Active Ingredient & Concentration:
325mg Aspirin/Tablet

Product Form:
Tablet

Other Ingredients:
Microcrystalline Cellulose, Potato Starch.

EXCEDRIN, ASPIRIN-FREE

Manufacturer: Bristol-Myers Products

Active Ingredients & Concentration:	Product Form:
500mg Acetaminophen +65mg Caffeine/Caplet	Caplet

Other Ingredients:

Benzoic Acid, Carnauba Wax, Corn Starch, Croscarmellose Sodium, Colors, Hydroxypropyl Methylcellulose, Magnesium Stearate, Methylparaben, Microcrystalline Cellulose, Propylparaben, Saccharin Sodium, Simethicone Emulsion, Stearic Acid, Titanium Dioxide. May also contain Erythorbic Acid, Mineral Oil, Polyethylene Glycol, Polysorbate 20, Polysorbate 80, Povidone, Propylene Glycol, Sorbitan Monolaurate.

EXCEDRIN, EXTRA STRENGTH

Manufacturer: Bristol-Myers Products

Active Ingredients & Concentration:	Product Forms:
250mg Acetaminophen + 250mg Aspirin + 65mg Caffeine/Caplet	Caplet Tablet

Other Ingredients:

Both tablets and caplets contain Benzoic Acid, FD&C Blue No.1, Hydroxypropyl Methylcellulose, Povidone, Propylene Glycol, Saccharin Sodium, Simethicone Emulsion, Sorbitan Monolaurate, Stearic Acid, Titanium Dioxide. May also contain Carnauba Wax, Hydroxypropylcellulose, Microcrystalline Cellulose, Mineral Oil, Polysorbate 20.

EXCEDRIN, SINUS

Manufacturer: Bristol-Myers Products

Active Ingredients & Concentration:	Product Forms:
500mg Acetaminophen + 30mg Pseudoephedrine Hydrochloride/ Tablet or Caplet	Caplet Tablet

Other Ingredients:

Corn Starch, D&C Yellow No. 10 Lake, FD&C Red No. 40 Lake, Hydroxypropyl Methylcellulose, Mineral Oil, Polysorbate 20, Povidone, Propylene Glycol, Simethicone Emulsion, Sorbitan Monolaurate, Stearic Acid, Titanium Dioxide. May also contain Benzoic Acid, Carnauba Wax.

HALTRAN

Manufacturer: Roberts Pharmaceutical Corp.

Active Ingredient & Concentration:	Product Form:
200mg Ibuprophen/Tablet	Tablet

Other Ingredients:

Carnauba Wax, Corn Starch, Hydroxypropyl Methylcellulose, Propylene Glycol, Silicon Dioxide, Pregelatinized Starch, Stearic Acid, Titanium Dioxide.

LIQUIPRIN

Manufacturer: Menley & James Laboratories, Inc.

Active Ingredient & Concentration:	Product Form:
80mg Acetaminophen/1.66ml	Liquid

Other Ingredients:

Flavors, Citric Acid, D&C Red No. 33, Dextrose, FD&C Red No. 40, Fructose, Glycerin, Methylparaben, Polyethylene Glycol, Propylene Glycol, Propylparaben, Sodium Citrate, Sodium Gluconate, Sucrose, Water.

MIDOL

Manufacturer: Sterling Winthrop, Inc.

Active Ingredients & Concentration:
325mg Acetaminophen
+ 12.5mg Pyrilamine Maleate/Caplet

Product Form:
Caplet

Other Ingredients:
Croscarmellose Sodium, Hydroxypropyl Methylcellulose, Magnesium Stearate, Microcrystalline Cellulose, Pregelatinized Starch, Triacetin.

MIDOL PMS

Manufacturer: Sterling Winthrop, Inc.

Active Ingredients & Concentration:
500mg Acetaminophen
+ 15mg Pyrilamine Maleate
+ 25mg Pamabrom/Caplet

Product Form:
Caplet

Other Ingredients:
Croscarmellose Sodium, Hydroxypropyl Methylcellulose, Magnesium Stearate, Microcrystalline Cellulose, Pregelatinized Starch, Triacetin.

MIDOL IB

Manufacturer: Sterling Winthrop, Inc.

Active Ingredient & Concentration:
200mg Ibuprophen/Tablet

Product Form:
Tablet

Other Ingredients:
Calcium Phosphate, Cellulose, Magnesium Stearate, Silicon Dioxide, Sodium Lauryl Sulfate, Sodium Starch Glycolate, Stearic Acid, Titanium Dioxide.

MIDOL, MAXIMUM STRENGTH

Manufacturer: Sterling Winthrop, Inc.

Active Ingredients & Concentration:
500mg Acetaminophen
+ 15mg Pyrilamine Maleate
+ 60mg Caffeine/Caplet

Product Form:
Caplet

Other Ingredients:
Croscarmellose Sodium, Hydroxypropyl Methylcellulose, Magnesium Stearate, Microcrystalline Cellulose, Pregelatinized Starch, Triacetin.

MIDOL, TEEN FORMULA

Manufacturer: Sterling Winthrop, Inc.

Active Ingredients & Concentration:
400mg Acetaminophen
+ 25mg Pamabrom/Caplet

Product Form:
Caplet

Other Ingredients:
Croscarmellose Sodium, Hydroxypropyl Methylcellulose, Magnesium Stearate, Microcrystalline Cellulose, Pregelatinized Starch, Triacetin.

MOBIGESIC

Manufacturer: B.F. Ascher & Company, Inc.

Active Ingredients & Concentration:
325mg Magnesium Salicylate
+ 30mg Phenyltoloxamine Citrate/Tablet

Product Form:
Tablet

Other Ingredients:
Microcrystalline Cellulose, Magnesium Stearate, Colloidal Silicon Dioxide.

MOTRIN IB

Manufacturer: The Upjohn Company

Active Ingredient & Concentration:
200mg Ibuprophen/Tablet or Caplet

Product Forms:
Tablet
Caplet

Other Ingredients:
Carnauba Wax, Corn Starch, Hydroxypropyl Methylcellulose, Propylene Glycol, Silicon Dioxide, Pregelatinized Starch, Stearic Acid, Titanium Dioxide.

NORWICH, REGULAR & MAXIMUM STRENGTH

Manufacturer: Chattem Consumer Products

Active Ingredient & Concentration:
Regular Strength: 325mg Aspirin/Tablet
Maximum Strength: 500mg Aspirin/Tablet

Product Form:
Tablet

Other Ingredients:
Hydroxypropyl Methylcellulose, Polyethylene Glycol, Starch.

NORWICH, ENTERIC

Manufacturer: Chattem Consumer Products

Active Ingredients & Concentration:
Regular Strength: 325mg Aspirin/Tablet
Maximum Strength: 500mg Aspirin/Tablet

Product Form:
Tablet

Other Ingredients:
Dioctyl Sodium Sulfosuccinate, Hydroxypropyl Methylcellulose, Hydroxypropylcellulose, Iron Oxide, Microcrystalline Cellulose, Pharmaceutical Glaze, Polyethylene Glycol, Polyvinyl Acetate Phthalate, Pregelatinized Starch, Silicon Dioxide, Stearic Acid, Talc, Titanium Dioxide, Triethyl Citrate, Colors.

NUPRIN

Manufacturer: Bristol-Myers Products

Active Ingredient & Concentration:
200mg Ibuprofen/Tablet or Caplet

Product Forms:
Tablet
Caplet

Other Ingredients:
Carnauba Wax, Corn Starch, D&C Yellow No. 10, FD&C Yellow No. 6, Hydroxypropyl Methylcellulose, Propylene Glycol, Silicon Dioxide, Stearic Acid, Titanium Dioxide.

P-A-C

Manufacturer: Roberts Pharmaceutical Corp.

Active Ingredients & Concentration:
400mg Aspirin + 32mg Caffeine/Tablet

Product Form:
Tablet

Other Ingredients:
Cellulose, Corn Starch, Croscarmellose Sodium, FD&C Blue No. 2, FD&C Yellow No. 5, Sucrose.

PAMPRIN

Manufacturer: Chattem Consumer Products

Active Ingredients & Concentration:
250mg Acetaminophen
+ 250mg Magnesium Salicylate
+ 25mg Pamabrom/Caplet

Product Form:
Caplet

Other Ingredients:
Hydroxypropyl Methylcellulose, Magnesium Stearate,, Microcrystalline Cellulose, Propylene Glycol, Silicon Dioxide, Starch, Stearic Acid, Titanium Dioxide, Color, Cellulose.

PAMPRIN, MAXIMUM STRENGTH

Manufacturer: Chattem Consumer Products

Active Ingredients & Concentration:
500mg Acetaminophen
+ 15mg Pyrilamine Maleate
+ 25mg Pamabrom/Caplet or Tablet

Product Form:
Caplet
Tablet

Other Ingredients:
Crospovidone, Magnesium Stearate, Povidone, Pregelatinized Starch, Sodium Starch Glycolate, Stearic Acid.

PANADOL, CHILDREN'S

Manufacturer: Sterling Winthrop, Inc.

Active Ingredients & Concentration:
Tablets: 80mg Acetaminophen/Tablet
Liquid: 160mg Acetaminophen/5ml
Drops: 80mg Acetaminophen/0.8ml

Product Forms:
Tablet, Ch.
Liquid
Drops

Other Ingredients:
Tablets contain FD&C Red No. 28, FD&C Red No. 40, Flavor, Mannitol, Saccharin Sodium, Starch, Stearic Acid and other ingredients.
Liquid contains Benzoic Acid, FD&C Red No. 40, Flavor, Glycerin, Polyethylene Glycol, Potassium Sorbate, Propylene Glycol, Purified Water, Saccharin Sodium, Sorbitol, May also contain Sodium Chloride or Sodium Hydroxide.
Drops contain Citric Acid, FD&C Red No. 40, Flavors, Glycerin, Parabens, Polyethylene Glycol, Propylene Glycol, Purified Water, Saccharin Sodium, Sodium Chloride, Sodium Citrate.

PANADOL, JUNIOR STRENGTH

Manufacturer: Sterling Winthrop, Inc.

Active Ingredient & Concentration:
160mg Acetaminophen/Caplet

Product Form:
Caplet

Other Ingredients:
Hydroxypropyl Methylcellulose, Potassium Sorbate, Povidone, Pregelatinized Starch, Starch, Stearic Acid, Talc, Triacetin.

PANADOL, MAXIMUM STRENGTH

Manufacturer: Sterling Winthrop, Inc.

Active Ingredient & Concentration:
500mg Acetaminophen/Caplet or Tablet

Product Forms:
Caplet
Tablet

Other Ingredients:
Hydroxypropyl Methylcellulose, Potassium Sorbate, Povidione, Pregelatinized Starch, Starch, Stearic Acid, Talc, Triacetin.

ST. JOSEPH, ADULT

Manufacturer: Schering-Plough HealthCare Products

Active Ingredient & Concentration:
81mg Aspirin/Tablet

Product Form:
Tablet, Ch.

Other Ingredients:
D&C Yellow No. 10 Aluminum Lake, FD&C Yellow No. 6 Aluminum Lake, Flavor, Hydrogenated Vegetable Oil, Maltodextrin, Mannitol, Saccharin, Corn Starch.

ST. JOSEPH, CHILDREN'S

Manufacturer: Schering-Plough HealthCare Products

Active Ingredient & Concentration:
80mg Acetaminophen/Tablet

Product Form:
Tablet, Ch.

Other Ingredients:
Cellulose, D&C Red No. 7 Calcium Lake, D&C Red No. 30 Aluminum Lake, Flavor, Mannitol, Silicon Dioxide, Saccharin Sodium, Zinc Stearate.

TYLENOL, REGULAR STRENGTH

Manufacturer: McNeil Consumer Products Company

Active Ingredient & Concentration:
325mg Acetaminophen/Tablet or Caplet

Product Forms:
Tablet
Caplet

Other Ingredients:
Tablets contain Magnesium Stearate, Cellulose, Sodium Starch Glycolate, Starch.
Caplets contain Cellulose, Hydroxypropyl Methylcellulose, Magnesium Stearate, Red No. 40, Polyethylene Glycol, Sodium Starch Glycolate, Starch.

TYLENOL, EXTRA STRENGTH

Manufacturer: McNeil Consumer Products Company

Active Ingredient & Concentration:
500mg Acetaminophen/Tablet, Caplet, Gelcap or 15ml of liquid

Product Forms:
Tablet
Caplet
Gelcap
Liquid

Other Ingredients:
Tablets contain Magnesium Stearate, Cellulose, Sodium Starch Glycolate, Starch.
Caplets contain Cellulose, Hydroxypropyl Methylcellulose, Magnesium Stearate, Red No. 40, Polyethylene Glycol, Sodium Starch Glycolate, Starch.
Gelcaps contain Benzyl Alcohol, Butylparaben, Castor Oil, Cellulose, Edetate Calcium Disodium, Gelatin, Hydroxypropyl Methylcellulose, Magnesium Stearate, Methylparaben, Propylparaben, Sodium Lauryl Sulfate, Sodium Propionate, Sodium Starch Glycolate, Starch, Titanium Dioxide, Colors.
Liquid contains Alcohol (7%), Citric Acid, Flavors, Glycerin, Polyethylene Glycol, Purified Water, Sodium Benzoate, Sorbitol, Sucrose, Colors.

TYLENOL, PAIN RELIEVER PLUS ANTACID

Manufacturer: McNeil Consumer Products Company

Active Ingredients & Concentration:	Product Form:
500mg Acetaminophen + 250mg Calcium Carbonate	Caplet

Other Ingredients:
Acacia, Cellulose, Corn Starch, Croscarmellose Sodium, Hydroxypropyl Methylcellulose, Magnesium Stearate, Maltodextrin, Propylene Glycol, Sodium Starch Glycolate, Titanium Dioxide, Triacetin, Colors.

TYLENOL PM, EXTRA STRENGTH

Manufacturer: McNeil Consumer Products Company

Active Ingredients & Concentration:	Product Forms:
500mg Acetaminophen + 25mg Diphenhydramine Hydrochloride/ Caplet, Gelcap or Tablet	Caplet Gelcap Tablet

Other Ingredients:
Tablets contain Cellulose, Colloidal Silicon Dioxide, Corn Starch, Sodium Citrate, Sodium Starch Glycolate, Stearic Acid, Colors.
Caplets contain Cellulose, Colloidal Silicon Dioxide, Corn Starch, Hydroxypropyl Methylcellulose, Polyethylene Glycol, Sodium Citrate, Sodium Starch Glycolate, Stearic Acid, Titanium Dioxide, Colors.
Gelcaps contain Benzyl Alcohol, Butylparaben, Castor Oil, Cellulose, Corn Starch, Edetate Calcium Disodium, Gelatin, Hydroxypropyl Methylcellulose, Magnesium Stearate, Methylparaben, Propylparaben, Sodium Lauryl Sulfate, Sodium Citrate, Sodium Propionate, Sodium Starch Glycolate, Titanium Dioxide, Colors.

TYLENOL, CHILDREN'S

Manufacturer: McNeil Consumer Products Company

Active Ingredients & Concentration:	Product Forms:
Drops: 80mg Acetaminophen/ 0.8ml of Liquid	Tablet, Ch.
Liquid: 160mg Acetaminophen/ 5ml of Suspension	Drops
Tablet: 80mg Acetaminophen/Tablet	Liquid

Other Ingredients:

Tablets contain Aspartame, Cellulose, Citric Acid, Ethylcellulose, Flavors, Hydroxypropyl Methylcellulose, Mannitol, Starch, Magnesium Stearate, Colors.

Suspension drops contain Butylparaben, Cellulose, Citric Acid, Corn Syrup, Flavors, Glycerin, Purified Water, Sodium Benzoate, Sorbitol, Xanthan Gum, FD&C Red No. 33, FD&C Blue No.1.

Suspension liquid contains Butylparaben, Cellulose, Citric Acid, Corn Syrup, Flavors, Glycerin, Purified Water, Sodium Benzoate, Sorbitol, Xanthan Gum, FD&C Red No. 40.

TYLENOL, JUNIOR STRENGTH

Manufacturer: McNeil Consumer Products Company

Active Ingredient & Concentration:	Product Forms:
160mg Acetaminophen/Tablet or Caplet	Tablet, Ch.
	Caplet

Other Ingredients:

Tablets contain Aspartame, Cellulose, Citric Acid, Ethylcellulose, Flavors, Magnesium Stearate, Mannitol, Starch, Blue No. 2, Red No. 7.

Caplets contain Cellulose, Ethylcellulose, Magnesium Stearate, Sodium Lauryl Sulfate, Sodium Starch Glycolate, Starch.

URSINUS

Manufacturer: Sandoz Pharmaceuticals/Consumer Division

Active Ingredients & Concentration: 325mg Aspirin + 30mg Pseudoephedrine Hydrochloride/Tablet

Product Form: Tablet

Other Ingredients: Calcium Stearate, Lactose, Microcrystalline Cellulose, Pregelatinized Starch, Sodium Starch, Colors.

VANQUISH

Manufacturer: Sterling Winthrop, Inc.

Active Ingredients & Concentration: 227mg Aspirin + 194mg Acetaminophen + 33mg Caffeine + 25mg Dried Aluminum Hydroxide Gel + 50mg Magnesium Hydroxide/Caplet

Product Form: Caplet

Other Ingredients: Acacia, Colloidal Silicon Dioxide, Hydrogenated Vegetable Oil, Microcrystalline Cellulose, Powdered Cellulose, Sodium Lauryl Sulfate, Starch, Talc.

PART II

SUPPLIER'S FORMULATIONS

INTERNAL ANALGESIC

Supplier: American Colloid Company

	Ingredient:	Wt.%
	Magnesium Aluminum Silicate (**Magnabrite K**)	1.50
A	Methylcellulose	0.15
	Water	91.00
	Saccharin Sodium	0.05
B	Bismuth Subsalicylate	3.00
	Salicylic Acid	0.40
	Sodium Salicylate	1.00
C	Ethanol	2.90
	Preservative, Color, Flavor	q.s.
		100.00

Mixing Procedure:

Dry blend the first two ingredients in Part A and add them to the water slowly, agitating continually until smooth. Add the Part B ingredients to Part A slowly with mixing until uniform. Add Part C and continue mixing until smooth.

Author's note: Although the name of this formula is "Internal Analgesic", its composition suggests it could also be classified as an antidiarrheal product because it contains 3% Bismuth Subsalicylate, a well known antidiarrheal active ingredient.

HIGH-DOSE ACETAMINOPHEN TABLET

Supplier: Aqualon Company

	Ingredient:	Wt.%
A	Acetaminophen, USP	83.30
B	Lactose, NF	5.10
C	Calcium Sulfate	5.10
D	Hydroxypropylcellulose (**Klucel EXF**)	4.00
E	Croscarmellose Sodium, Type A, NF	2.00
F	Magnesium Stearate,NF	0.50
		100.00

Mixing Procedure:

Preblend Parts A, B and C in a low shear planetary mixer. The binders can be add dry or as a solution, When added dry, water is used to granulate. The wet granulation is dried in a hot air oven to 0.5-1.0% moisture. After drying, the granulation is milled followed by blending with Parts E and F. A rotary tablet press is used to form the tablets.

MEDIUM-DOSE ACETAMINOPHEN TABLET

Supplier: Aqualon Company

	Ingredient:	Wt.%
A	Acetaminophen, USP	50.00
B	Lactose, NF	22.25
C	Calcium Sulfate	22.25
D	Hydroxypropylcellulose, (**Klucel EF**)	3.00
E	Croscarmellose Sodium, Type A, NF	2.00
F	Magnesium Stearate, NF	0.50
		100.00

Mixing Procedure:

Preblend Parts A, B and C in a low shear planetary mixer. The binders can be add dry or as a solution, When added dry, water is used to granulate. The wet granulation is dried in a hot air oven to 0.5-1.0% moisture. After drying, the granulation is milled followed by blending with parts E and F. A rotary tablet press is used to form the tablets.

ACETAMINOPHEN INSTANT GRANULES

Supplier: BASF Corp.

	Ingredient:	Wt.%
	Acetaminophen Powder (< 100μm)	15.06
	Sugar (< 100μm)	38.56
A	Crospovidone (**Kollidon CL-M**)	15.06
	Aspartame	2.11
	Flavors	3.01
B	Povidone (**Kollidon 30**)	3.61
	Ethanol, 96%	22.59
		100.00

Mixing Procedure:

Prepare a solution of with the Part B ingredients. Granulate a mixture of the Part A ingredients using the Part B solution. (1.3gm = 250mg of Acetaminophen)

PARACETAMOL (ACETAMINOPHEN) INSTANT GRANULES

Supplier: BASF Corp.

	Ingredient:	Wt.%
	Paracetamol Powder	26.60
	Sorbitol	69.22
A	Poloxamer 407 (**Lutrol F 127**)	1.06
	Citric Acid	1.56
	Sodium Citrate	1.56
B	Ethanol, 96%	q.s.
		100.00

Mixing Procedure:

Granulate a mixture of the Part A ingredients using Part B and dry. Fill in packets. (1.88gm = 500mg Acetaminophen).

ACETAMINOPHEN ORAL SOLUTION (500mg/10ml)

Supplier: BASF Corp.

	Ingredient:	Wt.%
A	Acetaminophen	5.00
B	Sorbitol, Crystalline	5.00
C	Povidone (**Kollidon 25**)	20.00
D	Sodium Cyclamate	3.00
E	Propylene Glycol	20.00
F	Flavor	0.20
G	Glycerin	15.00
H	Water	31.80
		100.00

Mixing Procedure:

Dissolve Parts A through F in a mixture of Parts G and H at room temperature.

ACETYL SALICYLIC ACID TABLETS (400mg)

Supplier: BASF Corp.

Ingredient:	Wt.%
Acetylsalicylic Acid, Crystalline	80.00
Ludipress	19.80
Stearic Acid	0.20
	100.00

Mixing Procedure:

Mix all the ingredients, pass through a sieve and press with medium compressive force.

IBUPROFEN GEL

Supplier: BASF Corp.

Ingredient:	Wt.%
Ibuprofen	5.00
Propylene Glycol	20.00
Poloxamer 407 (**Lutrol F 127**)	22.00
Water	53.00
	100.00

Mixing Procedure:

Mix all the ingredients at 6°C while stirring. After mixing for 1 hour, allow the batch to warm to room temperature, where gelation occurs.

INDOMETHACIN TABLETS (50mg)

Supplier: BASF Corp.

	Ingredient:	Wt.%
A	Indomethacin	12.32
	Lactose, Monohydrate	73.89
B	Povidone (**Kollidon 30**)	2.46
	Water	7.39
C	Crospovidone (**Kollidon CL**)	2.96
	Silicon Dioxide	0.49
	Magnesium Stearate	0.49
		100.00

Mixing Procedure:

Granulate the mixture of Part A ingredients with a solution prepared form the Part B ingredients. Pass through a 0.8mm sieve. Add the Part C ingredients, mix and press into tablets with low compressive force. The tablets should weigh 380mg and have a diameter of 12mm.

ACETAMINOPHEN TABLET (500mg)

Supplier: FMC Corp.

	Ingredient	Wt.%
A	Acetaminophen, Granular	74.40
B	Microcrystalline Cellulose (**Avicel® PH-101**)	19.10
C	Lactose, USP Anhydrous	4.50
D	Fumed Silica	0.30
E	Croscarmellose Sodium (**Ac-Di-Sol®**)	1.00
F	Magnesium Stearate	0.70
		100.00

Mixing Procedure:

Screen Parts B through F through a 40 mesh sieve and Part A through 20 mesh. Blend 1/2 of Part A with Parts B, C, D and E in a V-shaped blender for 8 minutes. Add the remainder of Part A to the above mixture and blend for 5 minutes. Add Part F and blend 3 minutes. Tablet using a tablet press with 7/16" S.C. punches to weight of 672 mg.

ANALGESIC TABLET

Supplier: FMC Corp.

	Ingredient:	Wt.%
A	Aspirin, 40 mesh	31.65
B	Salicylamide	15.45
C	Acetaminophen, 40 mesh	15.45
D	Caffeine, Granular	9.95
E	Microcrystalline Cellulose (**Avicel PH-101**)	25.00
F	Stearic Acid USP XXX	2.00
G	Fumed Silica	0.50
		100.00

Mixing Procedure:

Screen all the ingredients through a 20 mesh sieve. Blend all the ingredients in a V-shaped blender for 20 minutes. Compress 615mg tablets using 5/8" S.C. tooling.

ASPIRIN/ACETAMINOPHEN TABLET

Supplier: FMC Corp.

	Ingredient:	Wt.%
A	Aspirin, 40 mesh	37.30
B	Acetaminophen, 40 mesh	37.30
C	Microcrystalline Cellulose (**Avicel PH-102**)	22.40
D	Croscarmellose Sodium (**Ac-Di-Sol**)	1.20
E	Stearic Acid, USP XXX	1.20
F	Fumed Silica	0.60
		100.00

Mixing Procedure:

Screen all the ingredients through a 20 mesh sieve. Blend all the ingredients in a V-shaped blender for 10 minutes. Compress tablets of 670mg to the desired hardness using appropriate tooling.

ASPIRIN, CAFFEINE AND ACETAMINOPHEN TABLET (500mg)

Supplier: FMC Corp.

	Ingredient:	Wt.%
A	Aspirin, 40 mesh	33.33
B	Acetaminophen, 20 mesh	37.04
C	Caffeine, Granular	4.44
D	Microcrystalline Cellulose (**Avicel PH-102**)	14.82
E	Lactose, Anhydrous	6.67
F	Croscarmellose Sodium (**Ac-Di-Sol**)	1.50
G	Fumed Silica	0.70
H	Stearic Acid, XXX*, USP	1.50
		100.00

Mixing Procedure:

Mix Parts A through F in a suitable blender. Pass the mixture through a mill using a 12 mesh screen with knives forward. Add Parts G and H and blend the milled mixture for 20 minutes in a V-shaped blender. Compress tablets of 675mg to the desired hardness using appropriate tooling.

ASPIRIN, TABLET (325mg)

Supplier: FMC Corp.

	Ingredient:	Wt.%
A	Aspirin, 40 mesh	81.25
B	Microcrystalline Cellulose (**Avicel PH-101**)	11.30
C	Pregelatinized Starch	4.35
D	Stearic Acid, XXX*, USP	0.45
E	Croscarmellose Sodium (**Ac-Di-Sol**)	2.00
F	Fumed Silica	0.65
		100.00

Mixing Procedure:

Screen 50% of Part A through a mill using 12 mesh screen with knives forward. Preblend Parts B through F with 25% of Part A, then pass the mixture through the mill. Pass the remaining 25% of Part A through the mill. Mix all the ingredients in a V-shaped blender for 10 minutes. Compress using 13/32" S.C. tooling.

---

* XXX is a common abbreviation for triple-pressed Stearic Acid.

BUFFERED ASPIRIN TABLET

Supplier: FMC Corp.

	Ingredient:	Wt.%
A	Aspirin, 40 mesh	52.00
B	Magnesium Hydroxide	12.00
C	Aluminum Hydroxide	12.00
D	Microcrystalline Cellulose (**Avicel PH-102**)	17.60
E	Croscarmellose Sodium (**Ac-Di-Sol**)	2.00
F	Stearic Acid, XXX, USP	2.00
G	Hydroxide Coatings	2.40
		100.00

Mixing Procedure:

Screen all the ingredients except Part G through a 40 mesh sieve. Blend Parts B and C in a V-shaped blender for 10 minutes. Coat the Parts B+C mixture using **Aquacoat** aqueous polymeric dispersion in a fluidized bed column, using a 10% by weight formula. Blend 50% of Part A with all of Parts D and E for 10 minutes in a V-shaped blender. Add the remainder of ingredient A and blend for 10 minutes. Blend the coated hydroxides with the mixture from the previous step for 10 minutes. Add Part F and blend for 7 minutes. Compress tablets of 625 mg to the desired hardness using appropriate tooling.

CHEWABLE ACETAMINOPHEN TABLET

Supplier: FMC Corp.

	Ingredient:	Wt.%
A	Acetaminophen, Granular	21.55
B	Microcrystalline Cellulose (**Avicel RC-591**)	10.00
C	Fructose	30.00
D	Xylitol	35.25
E	Aspartame	1.00
F	Flavor	0.20
G	Citric Acid	1.00
H	Fumed Silica	0.50
I	Magnesium Stearate	0.50
		100.00

Mixing Procedure:

Sieve Parts B, C, D, E, G and I, separately, through a 40 mesh screen. Sieve Part H through a 20 mesh screen. Sieve Part F through a 100 mesh screen. Blend Parts A, E and F in a V-shaped blender for 3 minutes. Add Parts B, C, D, G and H and blend for an additional 10 minutes. Add Part I and blend for an additional 5 minutes.

CHILDREN'S CHEWABLE ACETAMINOPHEN TABLET (125mg)

Supplier: FMC Corp.

	Ingredient:	Wt.%
A	Acetaminophen	50.00
B	Mannitol	9.92
C	Fructose	9.92
D	Xylitol	9.92
E	Microcrystalline Cellulose (**Avicel PH-102**)	18.00
F	Aspartame	1.00
G	Flavor	0.20
H	Color	0.04
I	Fumed Silica	0.50
J	Magnesium Stearate	0.50
		100.00

Mixing Procedure:

Mix Parts A through E in a suitable mixer. Mix Parts F, G and H. Geometrically dilute the second mixture with the first mixture. Add the remainder of the first mixture and mix. Pass this mixture through a mill using a 20 mesh screen with knives forward Transfer to a V-shaped blender and add Part I, Mix for 10 minutes. Add Part J, mix for a minimum time (equivalent to 5 minutes in a V-shaped mixer). Compress tablets of 250 mg to a hardness of 6-10 kg.

(Note: If efficient taste masking is desired, Acetaminophen can be coated with **Aquacoat** aqueous polymeric dispersion plasticized with DBS. 30% of Parts B, C, D and H can be dissolved in this coating dispersion. Coating level applied should be approximately 15%)

ENTERIC COATED ASPIRIN TABLET (325mg)

Supplier: FMC Corp.

	Ingredient:	Wt.%
A	Aspirin USP, 40 mesh	81.25
B	Microcrystalline Cellulose (**Avicel PH-101**)	11.30
C	Pregelatinized Starch	4.35
D	Stearic Acid, XXX, USP	0.45
E	Croscarmellose Sodium (**Ac-Di-Sol**)	2.00
F	Fumed Silica	0.65
		100.00

Mixing Procedure:

Weigh all ingredients, screen 50% of Part A through a mill using a 12 mesh screen with knives forward. Pre-blend Parts B through F with 25% of the Part A. Pass the mixture through the mill. Pass the remaining 25% of Part A through the mill. Mix all ingredients in a V-shaped blender for 10 minutes. Compress using 13/32" S.C. tooling. Coat with **Aquateric CAP** dispersion.

POWDERED ACETAMINOPHEN TABLET

Supplier: FMC Corp.

	Ingredient:	Wt.%
A	Acetaminophen, Powder	71.40
B	Microcrystalline Cellulose (**Avicel PH-200**)	14.40
C	Microcrystalline Cellulose (**Avicel PH-105**)	10.00
D	Stearic Acid, XXX, USP	2.00
E	Croscarmellose Sodium (**Ac-Di-Sol**)	1.70
F	Fumed Silica	0.50
		100.00

Mixing Procedure:

Screen all ingredients except Part F through a 40 mesh sieve. Screen Part F through a 20 mesh sieve. Blend 50% of Part A with all of Parts B, E and F. Mix for 10 minutes in a V-shaped blender. Add the remainder of Part A and blend for 5 minutes. Add Part D and blend for 5 minutes. Compress to 700mg tablets using 1/12" S.C. tooling.

ANALGESIC AND ANTIPYRETIC SUPPOSITORIES

Supplier: Gattefossé s.a.

	Ingredient:	Wt.%
A	Acetaminophen	25.00
B	**Suppocire AML**	75.00
		100.00

Mixing Procedure:
Heat Part B to 50°C. Allow Part B to cool to 45°C and add Part A while stirring with a turbine mixer. Continue mixing and cooling and pour into molds at 34°C.

ANALGESIC SUPPOSITORIES

Supplier: Gattefossé s.a.

	Ingredient:	Wt.%
A	Acetylsalicylic Acid	20.00
B	**Suppocire AM**	80.00
		100.00

Mixing Procedure:
Heat Part B to 50°C. Allow Part B to cool to 40°C and add Part A while stirring with a turbine mixer. Continue mixing and cooling and pour into molds at 35°C, using molds previously chilled to 0 to -5°C. Remove the suppositories from the molds after 7 minutes.

ACETAMINOPHEN TABLETS

Supplier: International Specialty Products

	Ingredient:	Wt.%
A	Acetaminophen	71.27
B	Sucrose	13.15
C	Polyvinylpyrrolidone (**Plasdone®**), 10% Solution in Alcohol	q.s.
D	Stearic Acid	1.32
E	Talc	3.29
F	Corn Starch	6.58
G	Alginic Acid	4.39
		100.00

Mixing Procedure:

Mix Parts A and B and moisten with Part C. Pass through a 14 mesh screen to granulate. Allow the mixture to air dry. When dry, size through a 20 mesh screen, add Parts E, F and G, and mix for 10 minutes in a tumble blender. Add Part D and mix for 5 minutes. Compress into tablets using 13/32" flat face punches.

ACETAMINOPHEN TABLET (325mg)

Supplier: R.T. Vanderbilt Company, Inc.

Ingredient:	Wt.%
Acetaminophen	43.33
Magnesium Aluminum Silicate (**Veegum F**)	4.00
Emdex	30.67
Microcrystalline Cellulose	17.33
Stearic Acid	1.34
Fumed Silica	3.33
	100.00

Mixing Procedure:

Blend all the ingredients, except the Stearic Acid, for 15 minutes. Add the Stearic Acid and blend for an additional 10 minutes. Compress into capsule-shaped tablets.

CHAPTER 16

LAXATIVES

The active ingredients in brand name laxatives can be classified into five groups, bulk forming agents, emollients, lubricants, saline compounds and stimulants. The most commonly used bulk forming agent is Psyllium Hydrophilic Mucilloid. Docusate Sodium, a.k.a. Dioctyl Sodium Sulfosuccinate, is a widely used emollient(surfactant). Mineral oil is the lubricant of choice and inorganic and organic magnesium salts are popular saline compounds. Some of the stimulants used are Cascara Sagrada, Aloe, Casanthranol and Phenolphthalein.

A wide variety of product forms are available including capsules, chewable tablets, liquids, powders, suppositories, suspensions and tablets.

PART I

BRAND NAME FORMULATIONS

AGORAL

Manufacturer: Parke-Davis

Active Ingredients & Concentration:
4.2gm Mineral Oil
+ 0.2gm Phenolphthalein/15ml

Product Form:
Liquid

Other Ingredients:
Acacia, Agar, Benzoic Acid, Egg Albumin, Flavors, Glycerin, Sodium Benzoate, Tragacanth, Citric Acid or Sodium Hydroxide, Water. Raspberry flavor also contains D&C Red No. 30 Lake, Saccharin Sodium.

CITRUGEL, ADULT & CHILDREN'S

Manufacturer: SmithKline Beecham, Inc.

Active Ingredients & Concentration:
Adult Strength: 2gm Methylcellulose/ 19gm of Powder
Children's Strength: 1gm Methylcellulose/ 9.5gm of Powder

Product Form: Powder

Other Ingredients:
Citric Acid, FD&C Yellow No. 6, Flavors, Potassium Citrate, Riboflavin, Sucrose and other ingredients.

CITRUGEL, SUGAR FREE, ADULT & CHILDREN'S

Manufacturer: SmithKline Beecham, Inc.

Active Ingredients & Concentration:
Adult Strength: 2gm Methylcellulose/ 10.2gm of Powder
Children's Strength: 1gm Methylcellulose/ 5.1gm of Powder

Product Form: Powder

Other Ingredients:
Aspartame, Dibasic Calcium Phosphate, FD&C Yellow No. 6, Malic Acid, Maltodextrin, Flavors, Potassium Citrate, Riboflavin.

CORRECTOL

Manufacturer: Schering-Plough HealthCare Products

Active Ingredients & Concentration:
100mg Docusate Sodium
65mg Phenolphthalein/Tablet

Product Form: Tablet

Other Ingredients:
Butylparaben, Calcium Gluconate, Calcium Sulfate, Carnauba Wax, D&C Red No. 7 Calcium Lake, Gelatin, Magnesium Stearate, Sugar, Talc, Titanium Dioxide, Wheat Flour, White Wax and other ingredients.

CORRECTOL, EXTRA GENTLE

Manufacturer: Schering-Plough HealthCare Products

Active Ingredient & Concentration:	Product Form:
100mg Docusate Sodium/Tablet	Tablet

Other Ingredients:
Colors, Gelatin, Glycerin, Polyethylene Glycol 400, Propylene Glycol, Sorbitol.

COLACE

Manufacturer: Apothecon

Active Ingredient & Concentration:	Product Form:
Capsules: 50 or 100mg Docusate Sodium/ Capsule	Capsule
Liquid: 1% Docusate Sodium	Liquid
Syrup: 20mg Docusate Sodium/5ml	Syrup

Other Ingredients:
Capsules contain Citric Acid, Colors, Nonporcine Gelatin, Edible Ink, Polyethylene Glycol, Propylene Glycol, Purified Water. Capsules with 100mg active also contain Titanium Dioxide.
Liquid contains Citric Acid, Color, Methylparaben, Poloxamer, Polyethylene Glycol, Propylene Glycol, Propylparaben, Sodium Citrate, Vanillin, Water.
Syrup contains Alcohol (1% max,), Citric Acid, Colors, Flavors, Menthol, Methylparaben, Poloxamer, Polyethylene Glycol, Propylparaben , Sodium Citrate, Sucrose, Purified Water,

---

DOXIDAN

Manufacturer: The Upjohn Company

Active Ingredients & Concentration:	Product Form:
65mg Yellow Phenolphthalein + 60mg Docusate Calcium/Gelcap	Gelcap

Other Ingredients:
Alcohol (up to 1.5%, w/w), Corn Oil, FD&C Blue No. 1, Red No. 40, Gelatin, Glycerin, Hydrogenated Vegetable Oil, Lecithin, Parabens, Sorbitol, Titanium Dioxide, Vegetable Shortening, Yellow Wax and other ingredients.

DIALOSE

Manufacturer: Johnson & Johnson • MERCK Consumer Pharmaceuticals Company

Active Ingredient & Concentration: 100mg Docusate Sodium/Tablet

Product Form: Tablet

Other Ingredients: Colloidal Silicon Dioxide, Colors, Flavors, Hydroxypropyl Methylcellulose, Magnesium Stearate, Microcrystalline Cellulose, Polyethylene Glycol, Polysorbate 80, Pregelatinized Starch, Propylene Glycol, Sodium Starch Glycolate, Titanium Dioxide.

DIALOSE, PLUS

Manufacturer: Johnson & Johnson • MERCK Consumer Pharmaceuticals Company

Active Ingredients & Concentration: 100mg Docusate Sodium + 65mg Yellow Phenolphthalein/Tablet

Product Form: Tablet

Other Ingredients: D&C Yellow No. 10, Dextrates, Dibasic Calcium Phosphate Dihydrate, FD&C Red No. 40, Flavors, Hydroxypropyl Methylcellulose, Magnesium Stearate, Microcrystalline Cellulose, Polydextrose, Polyethylene Glycol, Polysorbate 80, Sodium Starch Glycolate, Titanium Dioxide, Triacetin.

DULCOLAX

Manufacturer: CIBA Consumer Pharmaceuticals

Active Ingredients & Concentration:	Product Forms:
Tablet: 5mg Bisacodyl/Tablet	Tablet
Suppository: 10mg Bisacodyl/Unit	Suppository

Other Ingredients:

Tablets contain Acacia, Acetylated Monoglyceride, Carnauba Wax, Cellulose Acetate Phthalate, Corn Starch, Colors, Dibutyl Phthalate, Docusate Sodium, Gelatin, Glycerin, Iron Oxides, Kaolin, Lactose, Magnesium Stearate, Methylparaben, Pharmaceutical Glaze, Polyethylene Glycol, Povidone, Propylparaben, Sodium Benzoate, Sorbitan Monooleate, Sucrose, Talc, Titanium Dioxide, White Wax.

Suppositories contain Hydrogenated Vegetable Oil.

EFFER-SYLLIUM

Manufacturer: Johnson & Johnson • MERCK Consumer Pharmaceuticals Company

Active Ingredient & Concentration:	Product Form:
3gm Psyllium Hydrocolloid/7gm Powder	Powder

Other Ingredients:

Citric Acid, Ethyl Vanillin, Flavors, Potassium Bicarbonate, Potassium Citrate, Saccharin Calcium, Starch, Sucrose.

EX-LAX, REGULAR STRENGTH

Manufacturer: Sandoz Pharmaceuticals/Consumer Div.

Active Ingredient & Concentration:
90mg Yellow Phenolphthalein/Tablet or Pill

Product Forms:
Tablet
Pill

Other Ingredients:
Tablets contain Cocoa, Sugar, Hydrogenated Palm Kernel Oil, Lecithin. Nonfat Dry Milk, Vanillin. Pills contain Acacia, Alginic Acid, Carnauba Wax, Colloidal Silicon Dioxide, Dibasic Calcium Phosphate, Iron Oxides, Magnesium Stearate, Microcrystalline Cellulose, Sodium Benzoate, Sodium Lauryl Sulfate, Starch, Stearic Acid, Sucrose, Talc, Titanium Dioxide.

EX-LAX, MAXIMUM RELIEF

Manufacturer: Sandoz Pharmaceuticals/Consumer Div.

Active Ingredient & Concentration:
135mg Yellow Phenolphthalein/Pill

Product Form:
Pill

Other Ingredients:
Acacia, Alginic Acid, Color, Carnauba Wax, Colloidal Silicon Dioxide, Dibasic Calcium Phosphate, Magnesium Stearate, Microcrystalline Cellulose, Povidone, Sodium Benzoate, Sodium Lauryl Sulfate, Starch, Stearic Acid, Sucrose, Talc, Titanium Dioxide.

EX-LAX, EXTRA GENTLE

Manufacturer: Sandoz Pharmaceuticals/Consumer Div.

Active Ingredients & Concentration:
65mg Yellow Phenolphthalein
+ 75mg Docusate Sodium/Pill

Product Form:
Pill

Other Ingredients:
Acacia, Croscarmellose Sodium, Dibasic Calcium Phosphate, Colloidal Silicon Dioxide, Magnesium Stearate, Microcrystalline Cellulose, Color, Stearic Acid, Sucrose, Talc, Titanium Dioxide.

EX-LAX, GENTLE NATURE

Manufacturer: Sandoz Pharmaceuticals/Consumer Div.

Active Ingredient & Concentration:	Product Form:
20mg Sennosides/Pill	Pill

Other Ingredients:
Alginic, Acid, Colloidal Silicon Dioxide, Dibasic Calcium Phosphate, Magnesium Stearate, Microcrystalline Cellulose, Pregelatinized Starch, Sodium Lauryl Sulfate, Stearic Acid.

FEEN-A-MINT

Manufacturer: Schering-Plough HealthCare Products

Active Ingredients & Concentration:	Product Forms:
Gum: 97.2mg Yellow Phenolphthalein/ Tablet Tablet: 65mg Yellow Phenolphthalein + 100mg Docusate Sodium/Pill	Tablet, Ch. Gum

Other Ingredients:
Gum contains Acacia, BHA, Carnauba Wax, Corn Starch, Gelatin, Glucose, Glycerin Gum Base, Peppermint Oil, Sodium Benzoate, Sugar Water, White Wax,
Tablets contain Butylparaben, Calcium Gluconate, Calcium Sulfate, Carnauba Wax, Gelatin, Magnesium Stearate, Sugar, Talc, Titanium Dioxide, Wheat Flour, White Wax and other ingredients.

FIBERALL

Manufacturer: CIBA Consumer Pharmaceuticals

Active Ingredients & Concentration:
Tablet: 1gm Calcium Polycarbophil/Tablet
Wafer or Powder: 3.4gm Psyllium Hydrophilic Mucilloid/Wafer or 5gm of Powder

Product Forms:
Tablet, Ch.
Powder
Wafer

Other Ingredients:
Tablets contain Crospovidone, Dextrose, Flavors, Magnesium Stearate, Color.
Wafers contain Flavors, Baking Powder, Corn Syrup, Flour, Glycerin, Sugar, Vegetable Oil, Wheat Bran.
Powder contains Citric Acid, Flavor, Polysorbate 60, Wheat Bran. Orange flavor also contains Beta-Carotene, Saccharin, Color.

HALEY'S M-O

Manufacturer: Sterling Winthrop, Inc.

Active Ingredients & Concentration:
301mg Magnesium Hydroxide + 1.25ml Mineral Oil/5ml

Product Forms:
Suspension

Other Ingredients:
Purified water.

KASOF

Manufacturer: Roberts Pharmaceutical Corp.

Active Ingredient & Concentration:
240mg Docusate Potassium/Capsule

Product Form:
Capsule

Other Ingredients:
Colors, Gelatin, Glycerin, Methylparaben, Polyethylene Glycol, Propylparaben, Purified Water, Sorbitol.

KONSYL

Manufacturer: Konsyl Pharmaceuticals, Inc.

Active Ingredient & Concentration:	Product Form:
100% Psyllium Hydrophilic Mucilloid	Powder

Other Ingredients:
None.

KONSYL-D

Manufacturer: Konsyl Pharmaceuticals, Inc.

Active Ingredient & Concentration:	Product Form:
3.4gm Psyllium Hydrophilic Mucilloid/6.5gm of Powder	Powder

Other Ingredients:
Dextrose.

KONSYL, ORANGE

Manufacturer: Konsyl Pharmaceuticals, Inc.

Active Ingredient & Concentration:	Product Form:
3.4gm Psyllium Hydrophilic Mucilloid/12gm of Powder	Powder

Other Ingredients:
Sucrose, Citric Acid, FD&C Yellow No. 6, D&C Yellow No. 10, Flavor.

METAMUCIL

Manufacturer: Procter & Gamble Company

Active Ingredients & Concentration:	Product Form:
Regular Flavor: 3.4gm Psyllium Hydrophilic Mucilloid/7gm of Powder Orange Flavor: 3.4gm Psyllium Hydrophilic Mucilloid/11gm of Powder	Powder

Other Ingredients:
Regular flavor contains Dextrose. Orange flavor contains Citric Acid, FD&C Yellow No. 6, Flavor, Sucrose.

METAMUCIL, SUGAR-FREE

Manufacturer: Procter & Gamble Company

Active Ingredient & Concentration:
3.4gm Psyllium Hydrophilic Mucilloid/5.2 or 5.4gm of Powder

Product Form:
Powder

Other Ingredients:
Aspartame, Citric Acid Flavor, Phenylalanine, Potassium Bicarbonate, Silicon Dioxide, Sodium Bicarbonate. May also contain Calcium Carbonate, FD&C Yellow No. 6.

METAMUCIL, SUNRISE SMOOTH

Manufacturer: Procter & Gamble Company

Active Ingredient & Concentration:
3.4gm Psyllium Hydrophilic Mucilloid/12gm of Powder

Product Form:
Powder

Other Ingredients:
Citric Acid, D&C Yellow No. 10, FD&C Yellow No. 6, Flavor, Sucrose.

METAMUCIL, SUNRISE SMOOTH, SUGAR-FREE

Manufacturer: Procter & Gamble Company

Active Ingredient & Concentration:
3.4gm Psyllium Hydrophilic Mucilloid/5.8gm of Powder

Product Form:
Powder

Other Ingredients:
Aspartame, Citric Acid, D&C Yellow No. 10, FD&C Yellow No. 6, Flavor, Phenylalanine, Maltodextrin. Regular flavor contains only Citric Acid, Magnesium Sulfate and Maltodextrin.

NATURE'S REMEDY

Manufacturer: SmithKline Beecham, Inc.

Active Ingredient & Concentration:
150mg Cascara Sagrada + 100mg Aloe/Tablet

Product Form:
Tablet

Other Ingredients:
Calcium Stearate, Cellulose, Lactose, Coating, Colors.

PERDIEM FIBER

Manufacturer: Rhone-Poulenc Rorer Pharmaceuticals,Inc.

Active Ingredient & Concentration:
4gm Psyllium/6gm of Powder

Product Form:
Powder

Other Ingredients:
Acacia, Iron Oxides, Flavor, Paraffin, Sucrose, Talc, Titanium Dioxide, undisclosed forms of Sodium and Potassium.

PERI-COLACE

Manufacturer: Apothecon

Active Ingredient & Concentration:
Capsule: 30mg Casanthranol
+ 100mg Docusate Sodium/Capsule
Liquid: 30mg Casanthranol
+ 60mg Docusate Sodium/15ml

Product Form:
Capsule
Liquid

Other Ingredients:
Capsules contain D&C Red No. 33, FD&C Red No. 40, Gelatin, Edible Ink, Polyethylene Glycol, Propylene Glycol, Titanium Dioxide, Purified Water.
Liquid contains Alcohol (10% v/v), Citric Acid, Flavors, Methyl Salicylate, Methylparaben, Poloxamer, Polyethylene Glycol, Propylparaben, Sodium Citrate, Sorbitol, Sucrose, Purified Water.

PHILLIPS'

Manufacturer: Sterling Winthrop, Inc.

Active Ingredients & Concentration:
90mg Phenolphthalein
+ 83mg Docusate Sodium/Gelcap

Product Form:
Gelcap

Other Ingredients:
FD&C Blue No. 2, Gelatin, Glycerin, PEG 400, PEG 3350, Propylene Glycol, Sorbitol, Titanium Dioxide.

PHILLIPS' MILK OF MAGNESIA

Manufacturer: Sterling Winthrop, Inc.

Active Ingredients & Concentration:
Suspension: 400mg Magnesium Hydroxide/5ml
Tablet: 311mg Magnesium Hydroxide/Tablet

Product Forms:
Suspension
Tablet

Other Ingredients:
Original flavor suspension contains Purified Water.
Mint and cherry flavored suspensions contain Carboxymethylcellulose Sodium, Citric Acid, Flavor, Glycerin, Microcrystalline Cellulose, Propylene Glycol, Purified Water, Sorbitol, Sucrose, Xanthan Gum. May also contain D&C Red No. 28.
Tablets contain Flavor, Starch, Sucrose.

PHILLIPS' MILK OF MAGNESIA, CONCENTRATED

Manufacturer: Sterling Winthrop, Inc.

Active Ingredient & Concentration:
800mg Magnesium Hydroxide/5ml

Product Form:
Suspension

Other Ingredients:
Carboxymethylcellulose Sodium, Citric Acid, Flavor, Glycerin, Microcrystalline Cellulose, Propylene Glycol, Purified Water, Sorbitol, Sucrose, Xanthan Gum, D&C Red No. 28 or FD&C Yellow No. 6.

SERUTAN

Manufacturer: Menley & James Laboratories, Inc.

Active Ingredient & Concentration: 2.5gm Psyllium Hydrophilic Mucilloid/6.5gm

Product Form: Powder

Other Ingredients: Acacia, BHA, Calcium Propionate, Caprylic/Capric Triglyceride, Color, Carboxymethylcellulose Sodium, Corn Starch, Sugar, Magnesium Stearate, Oat Flour, Sodium Benzoate, Saccharin Sodium, Sucrose, Vegetable Oil, Wheat Germ.

SURFAK

Manufacturer: The Upjohn Company

Active Ingredient & Concentration: 240mg Docusate Calcium/Gelcap

Product Form: Gelcap

Other Ingredients: Alcohol (up to 3% w/w), Corn Oil, Colors Gelatin, Glycerin, Parabens, Sorbitol and other ingredients.

SURGEON'S CHOICE

Manufacturer: The Biopractic Group II, Inc.

Active Ingredient & Concentration: 3.4gm Psyllium Hydrophilic Mucilloid + 0.01gm Docusate Sodium/7.5gm of Powder

Product Form: Powder

Other Ingredients: Dextrose, Fructose, Gelatin, Flavors, Citric Acid, FD&C Red No. 40.

SYLLACT

Manufacturer: Wallace Laboratories

Active Ingredient & Concentration:
3.3gm Powdered Psyllium Seed Husks/ Teaspoonful

Product Form:
Powder

Other Ingredients:
Citric Acid, Dextrose, FD&C Red No. 40, Flavor, Potassium Sorbate, Propylparaben, Methylparaben, Saccharin Sodium.

PART II

SUPPLIER'S FORMULATIONS

YELLOW PHENOLPHTHALEIN LAXATIVE (90mg)

Supplier: FMC Corp.

	Ingredient:	Wt.%
A	Yellow Phenolphthalein	25.00
B	Microcrystalline Cellulose (**Avicel® PH-102**)	18.00
C	Dicalcium Phosphate	52.00
D	Croscarmellose Sodium (**Ac-Di-Sol® SD-711**)	1.00
E	Fumed Silica	1.00
F	Stearic Acid, XXX	2.00
G	Magnesium Stearate	1.00
		100.00

Mixing Procedure:

Screen Parts F and G through a 40 mesh sieve. Blend Parts A and E in a V-shaped blender for 3 minutes. Add Parts B and D to the blender and mix for 5 minutes. Add Part C to the blender and mix for 12 minutes. Add Part F to the blender and mix for 3 minutes. Add Part G to the blender and mix for 5 minutes. Tablet on a Manesty Express 20 using 3/8" flat bevel punches to a hardness of 10kg. Average tablet weight = 360mg.

LAXATIVE SUPPOSITORIES

Supplier: Gattefossé s.a.

	Ingredient:	Wt.%
A	Glycerin	10.00
B	**Suppocire BS2X**	90.00
		100.00

Mixing Procedure:

Heat Part B to 50°C. Cool Part B to 45°C and add Part A while stirring with a turbine mixer. Continue cooling and pour into molds (previously chilled to 15°C) at 33°C. Average weight 2gm.

CHAPTER 17

OPHTHALMIC PRODUCTS

The ophthalmic products included in this chapter range from relatively simple lubricating solutions and ointments, also referred to as "artificial tears" products, to more sophisticated products containing vasoconstrictors such as Naphazoline Hydrochloride, Oxymetazoline Hydrochloride or Tetrahydrozoline Hydrochloride.

The products are available as either liquid eye drops or ointments.

PART I

BRAND NAME FORMULATIONS

CELLUFRESH

Manufacturer: Allergan Pharmaceuticals

Active Ingredient & Concentration:
0.5% Carboxymethylcellulose Sodium

Product Form:
Liquid

Other Ingredients:
Calcium Chloride, Magnesium Chloride, Potassium Chloride, Purified Water, Sodium Chloride, Sodium Lactate. May also contain Sodium Hydroxide or Hydrochloric Acid to adjust pH.

CELLUVISC

Manufacturer: Allergan Pharmaceuticals

Active Ingredient & Concentration:	Product Form:
1% Carboxymethylcellulose Sodium	Liquid

Other Ingredients:
Calcium Chloride, Potassium Chloride, Purified Water, Sodium Chloride, Sodium Lactate.

CLEAR EYES

Manufacturer: Ross Laboratories

Active Ingredients & Concentration:	Product Form:
0.012% Naphazoline Hydrochloride + 0.2% Glycerin	Liquid

Other Ingredients:
Boric Acid, Purified Water, Sodium Borate, Edetate Disodium, Benzalkonium Chloride.

CLEAR EYES, ACR

Manufacturer: Ross Laboratories

Active Ingredients & Concentration:	Product Form:
0.012% Naphazoline Hydrochloride + 0.25% Zinc Sulfate + 0.2% Glycerin	Liquid

Other Ingredients:
Boric Acid, Purified Water, Sodium Chloride, Sodium Citrate, Edetate Disodium, Benzalkonium Chloride.

COLLYRIUM FRESH

Manufacturer: Wyeth-Ayerst Laboratories

Active Ingredients & Concentration:	Product Form:
0.05% Tetrahydrozoline Hydrochloride + 1.0% Glycerin	Liquid

Other Ingredients:
Benzalkonium Chloride, 0.1% Edetate Disodium, Boric Acid, Hydrochloric Acid, Sodium Borate.

LACRI-LUBE

Manufacturer: Allergan Pharmaceuticals

Active Ingredient & Concentration:	Product Form:
57.3% White Petrolatum + 42.5% Mineral Oil	Ointment

Other Ingredients:
Lanolin Oils.

MURINE

Manufacturer: Ross Laboratories

Active Ingredients & Concentration:	Product Form:
0.5% Polyvinyl Alcohol + 0.6% Povidone	Liquid

Other Ingredients:
Benzalkonium Chloride, Dextrose, Edetate Disodium, Potassium Chloride, Purified Water, Sodium Bicarbonate, Sodium Chloride, Sodium Citrate, Sodium Phosphate (mono- and dibasic).

MURINE, PLUS

Manufacturer: Ross Laboratories

Active Ingredients & Concentration:	Product Form:
0.05% Tetrahydrozoline Hydrochloride + 0.5% Polyvinyl Alcohol + 0.6% Povidone	Liquid

Other Ingredients:
Benzalkonium Chloride, Dextrose, Edetate Disodium, Potassium Chloride, Purified Water, Sodium Bicarbonate, Sodium Chloride, Sodium Citrate, Sodium Phosphate (mono- and dibasic).

REFRESH PM

Manufacturer: Allergan Pharmaceuticals

Active Ingredients & Concentration:	Product Form:
56.8% White Petrolatum + 41.5% Mineral Oil	Ointment

Other Ingredients:
Lanolin Alcohols, Purified Water, Sodium Chloride.

TEARS PLUS

Manufacturer: Allergan Pharmaceuticals

Active Ingredients & Concentration:	Product Form:
1.4% Polyvinyl Alcohol +0.6% Povidone	Liquid

Other Ingredients:
0.5% Chlorobutanol, Purified Water, Sodium Chloride. May also contain Sodium Hydroxide or Hydrochloric Acid.

VISINE

Manufacturer: Pfizer Consumer Health Care

Active Ingredient & Concentration:	Product Form:
0.05% Tetrahydrozoline Hydrochloride	Liquid

Other Ingredients:
Boric Acid, Sodium Borate, Sodium Chloride, Water, Benzalkonium Chloride, Edetate Disodium.

VISINE, A.C.

Manufacturer: Pfizer Consumer Health Care

Active Ingredients & Concentration: Product Form:

0.05% Tetrahydrozoline Hydrochloride + 0.25% Zinc Sulfate — Liquid

Other Ingredients:

Boric Acid, Sodium Chloride, Sodium Citrate, Purified Water, Benzalkonium Chloride, Edetate Disodium.

VISINE, EXTRA

Manufacturer: Pfizer Consumer Health Care

Active Ingredients & Concentration: Product Form:

0.05% Tetrahydrozoline Hydrochloride + 1.0% Polyethylene Glycol 400 — Liquid

Other Ingredients:

Boric Acid, Sodium Borate, Sodium Chloride, Water, Benzalkonium Chloride, Edetate Disodium.

VISINE, L.R.

Manufacturer: Pfizer Consumer Health Care

Active Ingredient & Concentration: Product Form:

0.025% Oxymetazoline Hydrochloride — Liquid

Other Ingredients:

Boric Acid, Sodium Borate, Sodium Chloride, Water, 0.01% Benzalkonium Chloride, 0.1% Edetate Disodium.

CHAPTER 18

SEDATIVES & STIMULANTS

The active ingredients in brand name sedatives include diphenhydramine salts and doxylamine salts. These may also be combined with analgesics such as Acetaminophen to provide product line extensions that provide sleep aids with a pain and/or fever reducers. Caffeine is the only stimulant currently used in brand name products included in this chapter.

The brand name products are available as tablets, caplets powders and liquids.

PART I

BRAND NAME FORMULATIONS

A. SEDATIVES

BAYER SELECT, NIGHT TIME PAIN RELIEF

Manufacturer: Sterling Winthrop, Inc.

Active Ingredients & Concentration:
500mg Acetaminophen
+ 25mg Diphenhydramine Hydrochloride/Caplet

Product Form:
Caplet

Other Ingredients:
FD&C Blue No. 1 and No. 2, Hydroxypropyl Methylcellulose, Polyethylene Glycol, Polysorbate 80, Povidone, Starch, Talc, Titanium Dioxide, Potassium Sorbate, Stearic Acid.

BUFFERIN, AF

Manufacturer: Bristol-Myers Products

Active Ingredients & Concentration:	Product Form:
500mg Acetaminophen + 38mg Diphenhydramine Citrate/Caplet	Caplet

Other Ingredients:
Benzoic Acid, Carnauba Wax, Corn Starch, Colors, Hydroxypropyl Methylcellulose, Methylparaben, Magnesium Stearate, Propylene Glycol, Propylparaben, Simethicone Emulsion, Stearic Acid, Titanium Dioxide.

DOAN'S, P.M.

Manufacturer: CIBA Consumer Pharmaceuticals

Active Ingredients & Concentration:	Product Form:
500mg Magnesium Salicylate + 25mg Diphenhydramine Hydrochloride/Caplet	Caplet

Other Ingredients:
Colloidal Silicon Dioxide, Dibasic Calcium Phosphate, Croscarmellose Sodium, Lactose, Microcrystalline Cellulose, Magnesium Stearate, Stearic Acid, Color.

EXCEDRIN, P.M.

Manufacturer: Bristol-Myers Products

Active Ingredients & Concentration:	Product Forms:
Tablet/Caplet: 500mg Acetaminophen + 38mg Diphenhydramine Citrate/Tablet or Caplet	Caplet
Liquid: 1000mg Acetaminophen + 50mg Diphenhydramine Hydrochloride/30ml	Tablet
	Liquid

Other Ingredients:
Tablets and Caplets contain Benzoic Acid, Carnauba Wax, Corn Starch, Colors, Hydroxypropyl Methylcellulose, Methylparaben, Magnesium Stearate, Propylene Glycol, Propylparaben, Simethicone Emulsion, Stearic Acid, Titanium Dioxide.
Liquid contains Alcohol (10% by vol.), Benzoic Acid, FD&C Blue No. 1, Flavor, Polyethylene Glycol, Povidone, Sodium Citrate, Sucrose, Water.

MIDOL PM

Manufacturer: Sterling Winthrop, Inc.

Active Ingredients & Concentration:
500mg Acetaminophen
+ 25mg Diphenhydramine Hydrochloride/Caplet

Product Form:
Caplet

Other Ingredients:
Colors, Hydroxypropyl Methylcellulose, Polyethylene Glycol, Potassium Sorbate, Povidone, Pregelatinized Starch, Stearic Acid, Talc, Titanium Dioxide.

MILES NERVINE

Manufacturer: Miles, Inc.

Active Ingredient & Concentration:
25mg Diphenhydramine Hydrochloride/Caplet

Product Form:
Caplet

Other Ingredients:
Calcium Phosphate Dibasic, Calcium Sulfate, Carboxymethylcellulose Sodium, Corn Starch, Magnesium Stearate, Microcrystalline Cellulose.

NYTOL, REGULAR AND MAXIMUM STRENGTH

Manufacturer: Block Drug Company, Inc.

Active Ingredients & Concentration:
Regular strength: 25mg Diphenhydramine Hydrochloride/Tablet
Maximum strength: 50mg Diphenhydramine Hydrochloride/Tablet

Product Form:
Tablet

Other Ingredients:
Corn Starch, Lactose, Microcrystalline Cellulose, Silica, Stearic Acid. May also contain FD&C Blue No. 1.

SLEEPINAL

Manufacturer: Thompson Medical Company, Inc.

Active Ingredient & Concentration:	Product Forms:
50mg Diphenhydramine Hydrochloride/ Capsule or Powder Packet	Powder Capsule

Other Ingredients:

Capsules contain FD&C Blue No.1, Gelatin, Lactose, Magnesium Stearate, Povidone, Talc.

Powder contains Aspartame, Citric Acid, Colloidal Silicon Dioxide, Colors, Flavors, Lactose, Polyethylene Glycol, Povidone, Sodium Chloride, Sodium Citrate.

SOMINEX

Manufacturer: SmithKline Beecham, Inc.

Active Ingredients & Concentration:	Product Forms:
Tablets: 25mg Diphenhydramine Hydrochloride/Tablet Caplets: 50mg Diphenhydramine Hydrochloride/Caplet	Tablet Caplet

Other Ingredients:

Tablets contain Dibasic Calcium Phosphate, Magnesium Stearate, Microcrystalline Cellulose, Silicon Dioxide, Starch, FD&C Blue No. 1.

Caplets contain Carnauba Wax, Crospovidone, Dibasic Calcium Phosphate, Hydroxypropyl Methylcellulose, Magnesium Stearate, Microcrystalline Cellulose, Polyethylene Glycol, Polysorbate 80, Silicon Dioxide, Starch, Titanium Dioxide, White Wax, FD&C Blue No.1.

SOMINEX, PAIN RELIEF FORMULA

Manufacturer: SmithKline Beecham, Inc.

Active Ingredient & Concentration:	Product Form:
500mg Acetaminophen +25mg Diphenhydramine Hydrochloride/ Tablet	Tablet

Other Ingredients:

Crospovidone, Povidone, Silicon Dioxide, Starch, Stearic Acid, FD&C Blue No. 1.

UNISOM, NIGHTTIME SLEEP AID

Manufacturer: Pfizer Consumer Health Care

Active Ingredient & Concentration:
25mg Doxylamine Succinate/Tablet

Product Form:
Tablet

Other Ingredients:
Dibasic Calcium Phosphate, FD&C Blue No. 1 Aluminum Lake, Magnesium Stearate, Microcrystalline Cellulose, Sodium Starch Glycolate.

UNISOM, WITH PAIN RELIEF

Manufacturer: Pfizer Consumer Health Care

Active Ingredient & Concentration:
650mg Acetaminophen
+ 50mg Diphenhydramine Hydrochloride/
Tablet

Product Form:
Tablet

Other Ingredients:
Corn Starch, FD&C Blue No. 1 Aluminum Lake, FD&C Blue No. 2 Aluminum Lake, Hydroxypropyl Methylcellulose, Magnesium Stearate, Polyethylene Glycol, Polysorbate 80, Povidone, Stearic Acid, Titanium Dioxide.

B. STIMULANTS

NO DOZ

Manufacturer: Bristol-Myers Products

Active Ingredient & Concentration:
100mg Caffeine/Tablet

Product Form:
Tablet

Other Ingredients:
Corn Starch, Mannitol, Flavors, Microcrystalline Cellulose, Stearic Acid, Sucrose.

NO DOZ, MAXIMUM STRENGTH

Manufacturer: Bristol-Myers Products

Active Ingredient & Concentration:	Product Form:
200mg Caffeine/Caplet	Caplet

Other Ingredients:

Benzoic Acid, Corn Starch, FD&C Blue No. 1, Flavors, Hydroxypropyl Methylcellulose, Microcrystalline Cellulose, Propylene Glycol, Simethicone Emulsion, Stearic Acid Sucrose, Titanium Dioxide. May also contain, Carnauba Wax, Mineral Oil, Polysorbate 20, Povidone, Sorbitan Monolaurate.

VIVARIN

Manufacturer: SmithKline Beecham, Inc.

Active Ingredient & Concentration:	Product Forms:
200mg Caffeine Alkaloid/ Tablet or Caplet	Tablet Caplet

Other Ingredients:

Tablets contain Dextrose, Magnesium Stearate, Microcrystalline Cellulose, Powdered Cellulose, Silicon Dioxide, Starch, FD&C Yellow No. 6, D&C Yellow No. 10.

Caplets Contain Carnauba Wax, Dextrose, Hydroxypropyl Methylcellulose, Magnesium Stearate, Microcrystalline Cellulose, Polyethylene Glycol, Polysorbate 80, Powdered Cellulose, Silicon Dioxide, Starch Titanium Dioxide, White Wax, FD&C Yellow No. 6, D&C Yellow No. 10.

PART II

SUPPLIER'S FORMULATIONS

A. SEDATIVES

DIPHENHYDRAMINE HYDROCHLORIDE TABLET (25mg)

Supplier: FMC Corp.

	Ingredient:	Wt.%
A	Diphenhydramine Hydrochloride	10.00
B	Microcrystalline Cellulose (**Avicel® PH-101**)	20.00
C	Lactose, Anhydrous	68.20
D	Croscarmellose Sodium (**Ac-Di-Sol® SD-711**)	1.00
E	Fumed Silica	0.30
F	Stearic Acid, XXX*	0.20
G	Magnesium Stearate	0.30
		100.00

Mixing Procedure:

Screen Parts A, F and G, separately, through a 40 mesh sieve. Blend Parts A, B, D and E in a V-shaped blender for 3 minutes. Add Part C to the blender and mix for 17 minutes. Add Part F to the blender and mix for 3 minutes. Add Part G to the blender and mix for 5 minutes. Tablet on a Manesty Express 20 using 3/8" flat bevel punches to a hardness of 5.0kg. Average tablet weight = 251mg.

* XXX is a common abbreviation for triple-pressed Stearic Acid.

PYRILAMINE MALEATE TABLETS (25mg)

Supplier: FMC Corp.

	Ingredient:	Wt.%
A	Pyrilamine Maleate, Powder	12.50
B	Microcrystalline Cellulose (**Avicel PH-101**)	17.00
C	Lactose, Anhydrous	68.40
D	Croscarmellose Sodium (**Ac-Di-Sol SD-711**)	1.00
E	Fumed Silica	0.35
F	Stearic Acid, XXX	0.25
G	Magnesium Stearate	0.50
		100.00

Mixing Procedure:

Screen Parts A, F and G, separately, through a 40 mesh sieve. Blend Parts A and C in a V-shaped blender for 3 minutes. Add Parts B, D and E to the blender and mix for 17 minutes. Add Part F to the blender and mix for 3 minutes. Add Part G to the blender and mix for 5 minutes. Tablet on a Manesty Express 20 using 5/16" standard concave punches to a hardness of 5.5kg. Average tablet weight = 200mg.

B. STIMULANTS

CAFFEINE TABLETS (150mg)

Supplier: FMC Corp.

	Ingredient:	Wt.%
A	Caffeine, Powder	48.39
B	Microcrystalline Cellulose (**Avicel PH-102**)	18.00
C	Lactose, Anhydrous	15.08
D	Di-Pac, Granular	16.13
E	Croscarmellose Sodium (**Ac-Di-Sol SD-711**)	1.00
F	Fumed Silica	0.50
G	Stearic Acid, XXX	0.25
H	Magnesium Stearate	0.25
I	Flavor	0.40
		100.00

Mixing Procedure:

Screen Parts A, G and H, separately, through a 40 mesh sieve. Blend Parts A through F and I in a V-shaped blender for 20 minutes. Add Part G to the blender and mix for 3 minutes. Add Part H to the blender and mix for 5 minutes. Tablet on a Manesty Express 20 using 3/8" flat bevel punches to a hardness of 7.0kg. Average tablet weight = 311mg.

CHAPTER 19

SUNSCREENS

Since there are literally hundreds of brand name sunscreen products available today, it is impractical to try to include all of them in this chapter. We have selected a cross-section of brand name products that have a Sun Protection Factor (SPF) of 15 or greater. Most contain active ingredients that are classified as Category I in the long-awaited, recently-published U.S. FDA Tentative Final Monograph on sunscreen actives[1].

These 20 Category I ingredients and their maximum permitted use levels are:

1. Aminobenzoic Acid up to 15%
2. Cinoxate up to 3%
3. Diethanolamine Methoxycinnamate up to 10%
4. Digalloyl Trioleate up to 5%
5. Dioxybenzone up to 3%
6. Ethyl 4-[bis(hydroxypropyl)]Aminobenzoate to 5%
7. Glyceryl Aminobenzoate up to 3%
8. Homosalate up to 15%
9. Lawsone up to 0.25%, w/Dihydroxy Acetone to 3%
10. Menthyl Anthranilate up to 5%
11. Octocrylene up to 10%
12. Octyl Methoxycinnamate up to 7.5%
13. Octyl Salicylate up to 5%
14. Oxybenzone up to 6%
15. Padimate O up to 8%
16. Phenylbenzimidazole Sulfonic Acid up to 4%
17. Red Petrolatum up to 100%
18. Sulisobenzone up to 10%
19. Titanium Dioxide up to 25%
20. Trolamine Salicylate up to 12%

Since most brand name sunscreen manufacturers do not indicate the concentration of the active ingredients, these cannot be presented herein, but the list above and the formulae in Part II can serve as a guide to the formulator.

This chapter also contains the abbreviation "SPF", Sun Protection Factor, which is the ratio of the UV energy required to produce a minimum erythema dose (MED) on sunscreen-protected skin divided by the UV energy required to produce a MED on unprotected skin[1].

PART I

BRAND NAME FORMULATIONS

ALMAY, PROTECTIVE FACE CREAM

Manufacturer: Almay, Inc.

Active Ingredients & SPF
Octyl Methoxycinnamate
+ Titanium Dioxide, SPF 15

Product Form:
Lotion

Other ingredients:
Glycerin, Water, Butylene Glycol, Propylene Glycol Dicaprylate/Dicaprate, Safflower Oil, Stearyl Alcohol, Cetyl Alcohol, Hydrogenated Coco-glycerides, Steareth-20, PEG-20 Methyl Glucose Sesquistearate, Barium Sulfate, Petrolatum, Sodium PCA, Glyceryl Stearate, Retinyl Palmitate, Ascorbyl Palmitate, Tocopheryl Acetate, Hydroxyethylcellulose, Talc, PEG-16 Soya Sterol, Xanthan Gum, Magnesium Aluminum Silicate, Citric Acid, Edetate Disodium, Iron Oxides.

BABY GARDE SUNBLOCK

Manufacturer: Eclipse Laboratories, Inc.

Active Ingredients & SPF
Octyl Methoxycinnamate
+ Oxybenzone + Octyl Salicylate
+ Titanium Dioxide, SPF 25

Product Form:
Lotion

Other Ingredients:
Cetyl Alcohol, Diazolidinyl Urea, Methylparaben, Mineral Oil, Phenyl Trimethicone, Propylparaben, Stearic Acid, Triethanolamine, Water and other ingredients.

BAIN DE SOLEIL, ALL DAY SUNBLOCK

Manufacturer: Procter & Gamble Company

Active Ingredients & SPF:
Octyl Methoxycinnamate
+ Oxybenzone + Titanium Dioxide
+ Octocrylene, SPF 15 or SPF 30

Product Form:
Lotion

Other Ingredients:
Water, PVP/Eicosene Copolymer, Isohexadecane, Butylene Glycol, Dimethicone, Cyclomethicone, Panthenol, Triethanolamine, Cetyl Palmitate, Glyceryl Tribehenate, Stearoxytrimethylsilane, Stearyl Alcohol, Tocopheryl Acetate, DEA-Cetyl Phosphate, Carbomer, Acrylates/C10-30 Alkyl Acrylate Crosspolymer, DMDM Hydantoin, Iodopropynyl Butylcarbamate, Edetate Disodium.

BAIN DE SOLEIL, ALL DAY, FOR KIDS

Manufacturer: Procter & Gamble Company

Active Ingredients & SPF:
Octyl Methoxycinnamate
+ Oxybenzone + Titanium Dioxide
+ Octocrylene, SPF 30

Product Form:
Lotion

Other Ingredients:
Water, PVP/Eicosene Copolymer, Isohexadecane, Butylene Glycol, Dimethicone, Cyclomethicone, Panthenol, Triethanolamine, Cetyl Palmitate, Glyceryl Tribehenate, Stearoxytrimethylsilane, Stearyl Alcohol, Tocopheryl Acetate, DEA-Cetyl Phosphate, Carbomer, Acrylates/C10-30 Alkyl Acrylate Crosspolymer, DMDM Hydantoin, Iodopropynyl Butylcarbamate, Edetate Disodium.

CHAP STICK, SUNBLOCK 15

Manufacturer: A. H. Robins Company, Inc.

Active Ingredients & SPF:
7% Padimate O + 3% Oxybenzone
SPF 15

Product Form:
Lotion

Other Ingredients:
Petrolatum, FD&C Yellow No. 5 Aluminum Lake, Paraffin, Mineral Oil, White Wax, Isopropyl Lanolate, Camphor, Lanolin, Isopropyl Myristate, Cetyl Alcohol, Carnauba Wax, Fragrance, Methylparaben, Propylparaben, Titanium Dioxide, D&C Red No. 6 Barium Lake.

CHAP STICK, PETROLEUM JELLY PLUS

Manufacturer: A. H. Robins Company, Inc.

Active Ingredients & SPF:
7% Padimate O + 3% Oxybenzone
SPF 15

Product Form:
Lotion

Other Ingredients:
White Petrolatum, Aloe, Lanolin, Phenoxyethanol, Methylparaben, Ethylparaben, Propylparaben, Butylparaben, BHT, Fragrance

---

CHILD GARDE, SUNBLOCK

Manufacturer: Eclipse Laboratories, Inc.

Active Ingredients & SPF:
Padimate O + Oxybenzone
+ Octyl Methoxycinnamate
SPF 30

Product Form:
Lotion

Other Ingredients:
Cetyl Alcohol, Diazolidinyl Urea, Methylparaben, Mineral Oil, Phenyl Trimethicone, Propylparaben, Stearic Acid, Triethanolamine, Water, Iron Oxides and other ingredients.

COPPERTONE, ALL DAY

Manufacturer: Schering-Plough HealthCare Products

Active Ingredients & SPF:
Octyl Methoxycinnamate
+ Octyl Salicylate
+ Homosalate + Oxybenzone
SPF 25

Product Form:
Lotion

Other Ingredients:
Water, Sorbitan Oleate, Sorbitol, Glyceryl Stearate SE, Stearic Acid, Hydrogenated Vegetable Oil, Octadecene/Maleic Anhydride Copolymer, Triethanolamine, Benzyl Alcohol, Fragrance, Dimethicone, Methylparaben, Imidazolidinyl Urea, Propylparaben, Tocopherol, Aloe Extract, Jojoba Oil, Carbomer, Edetate Disodium.

COPPERTONE, SPORT

Manufacturer: Schering-Plough HealthCare Products

Active Ingredients & SPF:
SPF 15: Oxybenzone
+ Octyl Methoxycinnamate
SPF 30: Oxybenzone
+ Octyl Methoxycinnamate + Octyl Salicylate

Product Form:
Lotion

Other Ingredients:
Water, PEG-8, Octyl Palmitate, Silica, DEA-Cetyl Phosphate, Acrylates/C10-30 Alkyl Acrylate Crosspolymer, Diazolidinyl Urea, Methylparaben, Aloe Extract, Jojoba Oil, Tocopheryl Acetate, Propylparaben, Fragrance, Edetate Disodium.

EUCERIN, DAILY FACIAL LOTION

Manufacturer: Beiersdorf, Inc.

Active Ingredients & SPF:
Octyl Methoxycinnamate
+ Phenylbenzimidazole Sulfonic Acid
+ Octyl Salicylate + Titanium Dioxide
SPF 20

Product Form:
Lotion

Other Ingredients:
Purified Water, Caprylic/Capric Triglyceride, Mineral Oil, Octyl Stearate, Cetearyl Alcohol, Glyceryl Stearate, Sodium Hydroxide, PEG-40 Castor Oil, Acrylamide/Sodium Acrylate Copolymer, Sodium Cetearyl Sulfate, Lanolin Alcohol, EDTA, Methylchloroisothiazolinone, Methylisothiazolinone.

JOHNSON & JOHNSON'S BABY SUNBLOCK

Manufacturer: Johnson & Johnson Consumer Products

Active Ingredients & SPF:
Octyl Methoxycinnamate
+ Octyl Salicylate + Oxybenzone
+ Titanium Dioxide
SPF 30

Product Form:
Lotion

Other Ingredients:
BHT, C12-15 Alcohols Benzoate, Carbomer, Cetyl Alcohol, DEA-Cetyl Phosphate, Dimethicone, Dioctyl Sodium Sulfosuccinate, Edetate Disodium, Fragrance, Glyceryl Dilaurate, Isopropyl Isostearate, PPG-2-Isodeceth-7-Carboxylate, Isostearic Acid, Propylene Glycol, Quaternium-15, Simethicone, Tocopheryl Acetate, Water and other ingredients.

NEUTROGENA, SUNBLOCK SPF 15

Manufacturer: Neutrogena Corp.

Active Ingredients & SPF:
Octyl Methoxycinnamate
+ Octyl Salicylate
+ Menthyl Anthranilate
SPF 15

Product Form:
Lotion

Other Ingredients:
Mineral Oil, Aluminum Starch Octenylsuccinate, Silica, PVP/Eicosene Copolymer, Phenyltrimethicone, Cyclomethicone, Tribehenin, Calcium Behenate, Titanium Dioxide, Dimethiconol, Propylparaben, C18-36 Acid Triglyceride.

NEUTROGENA, SUNBLOCK SPF 30

Manufacturer: Neutrogena Corp.

Active Ingredients & SPF:
Octyl Methoxycinnamate
+ Octocrylene
+ Menthyl Anthranilate
SPF 30

Product Form:
Lotion

Other Ingredients:
Mineral Oil, Aluminum Starch Octenylsuccinate, Silica, Phenyl Trimethicone, Tricontanyl PVP, Tribehenin(and)Calcium Behenate, Cyclomethicone(and)Dimethiconol, Stearyl Dimethicone, Zinc Oxide, Tocopheryl Acetate, BHT, C18-36 Acid Triglyceride.

OIL OF OLAY, DAILY UV PROTECTANT

Manufacturer: Procter & Gamble Company

Active Ingredients & SPF:
- Octyl Methoxycinnamate
- \+ Phenylbenzimidazole Sulfonic Acid
- SPF 15

Product Forms:
- Lotion
- Cream

Other Ingredients:

Lotion and cream contain Water, Isohexadecane, Butylene Glycol, Triethanolamine, Glycerin, Stearic Acid, Cetyl Alcohol, Cetyl Palmitate, DEA-Cetyl Phosphate, Aluminum Starch Octenylsuccinate, Titanium Dioxide, Imidazolidinyl Urea, Methylparaben, Propylparaben, Carbomer, Acrylates/C10-30 Alkyl Acrylate Crosspolymer, PEG-10 Soya Sterol, Edetate Disodium, Castor Oil, Fragrance, FD&C Red No. 4, FD&C Yellow No.5. Also available 100% color-free and fragrance-free.

PHOTOPLEX, BROAD SPECTRUM SUNSCREEN

Manufacturer: Allergan Pharmaceuticals

Active Ingredients & Concentration:
- 3.0% Avobenzone
- 7.0% Padimate O
- SPF 15

Product Form:
- Lotion

Other Ingredients:

Benzyl Alcohol, Carbomer 934P, Cetyl Esters Wax, Edetate Disodium, Glycerin, Imidazolidinyl Urea, Mineral Oil, Oleth-3 Phosphate, Purified Water, Stearyl Alcohol(and)Ceteareth-20, White Petrolatum. May also contain Sodium Hydroxide or Hydrochloric Acid.

PRESUN, ACTIVE 15 AND 30

Manufacturer: Bristol-Myers Products

Active Ingredients & SPF:
Octyl Methoxycinnamate
+ Octyl Salicylate + Oxybenzone
SPF 15 or SPF 30

Product Form:
Gel

Other Ingredients:
SPF 15 gel contains Alcohol SD40 (71.5%).
SPF 30 gel contains Alcohol SD40 (69%).
Both also contain Acrylates/t-Ocylpropenamide Copolymer, PPG-15 Stearyl Ether, Hydroxypropylcellulose.

PRESUN, FOR KIDS

Manufacturer: Bristol-Myers Products

Active Ingredients & SPF:
Octyl Methoxycinnamate
+ Octyl Salicylate + Oxybenzone
SPF 29

Product Form:
Lotion

Other Ingredients:
Water, Isopropyl Myristate, PG Dioctanoate, Isodecyl Neopentanoate, DEA-Cetyl Phosphate, PVP/Eicosene Copolymer, Stearic Acid, Cetyl Alcohol, Dimethicone, Diazolidinyl Urea, Carbomer 940, Triethanolamine, Methylchloroisothiazolinone, Methylisothiazolinone.

PRESUN 15

Manufacturer: Bristol-Myers Products

Active Ingredients & SPF:
Padimate O + Oxybenzone
SPF 15

Product Form:
Lotion

Other Ingredients:
Water, Petrolatum, Isopropyl Myristate, PG Dioctanoate, Isodecyl Neopentanoate, DEA-Cetyl Phosphate, PVP/Eicosene Copolymer, Stearic Acid, Cetyl Alcohol, Dimethicone, Diazolidinyl Urea, Carbomer 940, Triethanolamine, Methylchloroisothiazolinone, Methylisothiazolinone. May also contain Fragrance.

PRESUN 25

Manufacturer: Bristol-Myers Products

Active Ingredients & SPF:
Octyl Methoxycinnamate + Oxybenzone + Octyl Salicylate, SPF 25

Product Form:
Lotion

Other Ingredients:
Water, Isopropyl Myristate, Isodecyl Neopentanoate, PG Dioctanoate, Petrolatum, DEA-Cetyl Phosphate, PVP/Eicosene Copolymer, Stearic Acid, Cetyl Alcohol, Dimethicone, Diazolidinyl Urea, Carbomer 940, Triethanolamine, Methylchloroisothiazolinone, Methylisothiazolinone. May also contain Fragrance.

PRESUN 46

Manufacturer: Bristol-Myers Products

Active Ingredients & SPF:
Padimate O + Oxybenzone
SPF 46

Product Form:
Lotion

Other Ingredients:
Water, Isopropyl Myristate, Isodecyl Neopentanoate, PG Dioctanoate, DEA-Cetyl Phosphate, PVP/Eicosene Copolymer, Stearic Acid, Cetyl Alcohol, Dimethicone, Diazolidinyl Urea, Carbomer 940, Triethanolamine, Methylchloroisothiazolinone, Methylisothiazolinone. May also contain Fragrance.

PRESUN 15 AND 29, SENSITIVE SKIN

Manufacturer: Bristol-Myers Products

Active Ingredients & SPF:
Octyl Methoxycinnamate + Oxybenzone + Octyl Salicylate, SPF 15 or 29

Product Form:
Lotion

Other Ingredients:
Both SPF products contain Water, Isopropyl Myristate, Isodecyl Neopentanoate, PG Dioctanoate, DEA-Cetyl Phosphate, PVP/Eicosene Copolymer, Stearic Acid, Cetyl Alcohol, Dimethicone, Diazolidinyl Urea, Carbomer 940, Triethanolamine, Methylchloroisothiazolinone, Methylisothiazolinone.

SHADE, UVA GUARD

Manufacturer: Schering-Plough HealthCare Products

Active Ingredients & Concentration:
7.5% Octyl Methoxycinnamate
+3% Avobenzone
+ 3% Oxybenzone
SPF 15

Product Form:
Lotion

Other Ingredients:
Benzyl Alcohol, Carbomer 941, Dimethicone, Edetate Disodium, Glyceryl Stearate SE, Isopropyl Myristate, Methylparaben, Octadecene/Maleic Anhydride Copolymer, Propylparaben, Purified Water, Sorbitan Monooleate, Sorbitol, Stearic Acid, Trolamine.

SHADE, SUNBLOCK

Manufacturer: Schering-Plough HealthCare Products

Active Ingredients & SPF:
Octyl Methoxycinnamate
+ Oxybenzone + Octyl Salicylate
+ Octocrylene
SPF 45

Product Form:
Lotion

Other Ingredients:
Water, Sorbitan Isostearate, Sorbitol, Octadecene/Maleic Anhydride Copolymer, TEA, Stearic Acid, Benzyl Alcohol, Barium Sulfate, Dimethicone, Methylparaben, Imidazolidinyl Urea, Phenethyl Alcohol, Propylparaben, Carbomer 941, Edetate Disodium.

TOTAL ECLIPSE

Manufacturer: Eclipse Laboratories, Inc.

Active Ingredients & SPF:
Oxybenzone + Padimate O
SPF 15

Product Form:
Lotion

Other Ingredients:
Oleth-3 Phosphate, Glycerin, Benzyl Alcohol, Cetyl Esters, Carbomer 934, Imidazolidinyl Urea, Fragrance, Water, Sodium Hydroxide.

VASELINE, INTENSIVE CARE MOISTURIZING SUNBLOCK

Manufacturer: Chesebrough-Pond's, Inc.

Active Ingredients & SPF: Octyl Methoxycinnamate + Oxybenzone SPF 15

Product Form: Lotion

Other Ingredients: Water, Glycerin, Stearic Acid, Aloe Vera Gel, Fragrance, PVP/Eicosene Copolymer, Cyclomethicone, Triethanolamine, DEA-Cetyl Phosphate, Citronella Oil, Cetyl Alcohol, Petrolatum, Tocopherol Acetate, Magnesium Aluminum Silicate, Carbomer, Methylparaben, Propylparaben, Edetate Disodium, DMDM Hydantoin.

WATER BABIES, SUNBLOCK 15

Manufacturer: Tanning Research Laboratories, Inc.

Active Ingredient & SPF: Octyl Methoxycinnamate + Oxybenzone SPF 15

Product Form: Lotion

Other Ingredients: Water, Sorbitan Oleate, Sorbitol, Glyceryl Stearate SE, Hydrogenated Vegetable Oil, Isopropyl Myristate, Stearic Acid, Octadecene/Maleic Anhydride Copolymer, Benzyl Alcohol, Triethanolamine, Fragrance, Dimethicone, Methylparaben, Imidazolidinyl Urea, Propylparaben, Carbomer 941, Edetate Disodium.

WATER BABIES, SUNBLOCK 30

Manufacturer: Tanning Research Laboratories, Inc.

Active Ingredient & SPF:
Octyl Methoxycinnamate + Oxybenzone + Homosalate + Octyl Salicylate
SPF 30

Product Form:
Lotion

Other Ingredients:
Water, Sorbitan Oleate, Sorbitol, Glyceryl Stearate SE, Hydrogenated Vegetable Oil, Stearic Acid, Octadecene/Maleic Anhydride Copolymer, Benzyl Alcohol, Triethanolamine, Fragrance, Dimethicone, Methylparaben, Imidazolidinyl Urea, Propylparaben, Magnesium Aluminum Silicate, Tocopherol, Edetate Disodium.

WATER BABIES, SUNBLOCK 45

Manufacturer: Tanning Research Laboratories, Inc.

Active Ingredient & SPF:
Octyl Methoxycinnamate + Oxybenzone + Octyl Salicylate + Octocrylene
SPF 45

Product Form:
Lotion

Other Ingredients:
Water, Sorbitol, Stearic Acid, Sorbitan Isostearate, Barium Sulfate, Octadecene/Maleic Anhydride Copolymer, Benzyl Alcohol, Triethanolamine, Fragrance, Dimethicone, Methylparaben, Imidazolidinyl Urea, Propylparaben, Carbomer 941, Tocopherol, Edetate Disodium.

PART II

SUPPLIER'S FORMULATIONS

SUNSCREEN CREAM (SPF 15)

Supplier: BASF Corp.

	Ingredient:	Wt.%
A	Distilled Water	65.70
B	Carbomer 940	0.25
C	PEG-8 (**Pluracol E 400**)	3.00
D	Edetate Disodium	0.05
E	Cetearyl Octanoate (**Luvitol EHO**)	3.00
F	Ceteraeth-6(and)Stearyl Alcohol (**Cremophor A 6**)	1.50
G	Ceteareth-25 (**Cremophor A 25**)	2.50
H	Octocrylene (**Uvinul N 539 SG**)	10.00
I	Octyl Methoxycinnamate	7.50
J	Glyceryl Stearate(and)PEG-100 Stearate	2.00
K	Stearic Acid	2.00
L	Distilled Water	2.25
M	Tetrahydroxypropyl Ethylenediamine (**Neutrol TE**)	0.25
N	Fragrance, Preservative	q.s.
		100.00

Mixing Procedure:

Combine Parts A through D and heat to 75-80°C. Combine Parts E through K and heat to 75-80°C. Add the Part A-D mixture to the heated Part E-K mixture with stirring. When the batch is completely uniform, add combined Parts L and M and continue mixing and cooling until the batch is at 30-35°C. Add the Part N ingredients according to the manufacturer's suggestions and continue mixing until the batch reaches room temperature.

SUNSCREEN CREAM (SPF 20+)

Supplier: BASF Corp.

	Ingredient:	Wt.%
A	Distilled Water	61.60
B	Carbomer 940	0.25
C	PEG-8 (**Pluracol E 400**)	3.00
D	Edetate Disodium	0.05
E	Cetearyl Octanoate (**Luvitol EHO**)	3.00
F	Ceteraeth-6(and)Stearyl Alcohol (**Cremophor A 6**)	1.50
G	Cetearyl Alcohol	1.00
H	Oxybenzone (**Uvinul M 40**)	3.00
I	Octyl Methoxycinnamate	7.50
J	Octyl Salicylate	5.00
K	Glyceryl Stearate(and)PEG-100 Stearate	0.50
L	Stearic Acid	2.00
M	Cetearyl Alcohol	1.00
N	Distilled Water	10.00
O	Tetrahydroxypropyl Ethylenediamine (**Neutrol TE**)	0.60
P	Fragrance, Preservative	q.s.
		100.00

Mixing Procedure:

Combine Parts A through D and heat to 75-80°C. Combine Parts E through M and heat to 75-80°C. Add the heated Parts A-D to the heated Parts E-M with stirring. When the batch is completely uniform, add combined Parts N and O and continue mixing and cooling until the batch is at 30-35°C. Add Part P, according to the manufacturer's suggestions, and continue mixing until the batch reaches room temperature.

MOISTURIZING SUNTAN LOTION

Supplier: Calgon Corp.

	Ingredient:	Wt.%
A	Deionized Water	69.41
B	PPG-12 Buteth-16	4.00
	Carbomer 941	0.09
C	Triethanolamine	1.00
D	Polyquaternium 7 (**Merquat S**)	3.00
	Deionized Water	5.00
	Cetyl Alcohol	4.00
	Padimate O	4.00
	Mineral Oil, Light	2.00
	Isopropyl Palmitate	2.00
E	Acetylated Lanolin Alcohol	2.00
	Sorbitan Stearate	2.00
	Polysorbate 60	1.00
	Paraffin	0.20
F	Fragrance	0.30
	Preservative	q.s.
		100.00

Mixing Procedure:

Heat Part A to 85°C. Thoroughly mix the Part B ingredients and add the mixture to Part A with stirring. Add Part C at 85°C and continue mixing. Premix the Part D ingredients and add them to the above mixture with agitation, maintaining 80°C. In a separate vessel, mix and heat the Part E ingredients to 78°C. Slowly add Part E to the above mixture with good agitation. Begin cooling when uniform. Add the Part F ingredients at 45°C with agitation. Pour off at 30°C.

SUNSCREEN

Supplier: Calgon Corp.

	Ingredient:	Wt.%
	Xanthan Gum (**Keltrol CG**)	0.50
A	Deionized Water	69.35
	Sodium Chloride	0.15
	Glycerin	3.00
	Mineral oil	2.00
	Octyl Methoxycinnamate	7.50
	Oxybenzone	3.50
	Emulsifying Wax NF	3.00
B	C12-15 Alcohols Benzoate	2.00
	Titanium Dioxide(and)Octyl Palmitate	5.00
	Mineral Oil(and)Lanolin Alcohol	2.00
	Dimethicone, 100cs.	1.00
	Isostearyl Isopentanoate	1.00
C	Fragrance, Preservative	q.s.
		100.00

Mixing Procedure:

Mix **Keltrol CG** with Sodium Chloride and add this mixture to the Water and Glycerin. Mix until the Xanthan Gum is fully hydrated. Heat Parts A and B, separately, to 75°C. With continuous agitation, add Part B to Part A. Continue mixing while cooling to 40°C. Add Part C and mix until uniform.

SUNTAN LOTION

Supplier: Dow Chemical Company

	Ingredient:	Wt.%
A	Deionized Water	68.35
	Tetrasodium EDTA (**Versene Powder**)	0.05
	Carbomer 941	0.15
B	Glycerin	3.00
	Propylene Glycol	1.00
	Methylparaben	0.20
	Ethylparaben	0.15
C	Propylene Glycol	2.00
	Xanthan Gum	0.10
	Hydroxypropyl Methylcellulose (**Methocel 40-101**)	0.10
D	Mineral Oil	10.00
	Isopropyl Palmitate	2.00
	Glyceryl Monostearate	3.00
	Sorbitan Stearate	1.00
	Stearic Acid	2.00
	Dimethicone	0.50
	Padimate O	1.50
	Petrolatum	1.00
	Cetyl Alcohol	1.00
E	Deionized Water	1.00
	Triethanolamine	0.65
F	Deionized Water	1.00
	Quaternium-15 (**Dowicil 200**)	0.10
G	Fragrance	0.15
		100.00

Mixing Procedure:

Add The second and third ingredients in Part A to the Part A Water with rapid mixing until dissolved and heat Part A to 80°C. Mix the Part B ingredients and warm them to dissolve the Parabens. Add Part B to Part A while heating. Mix the Part C ingredients to fully wet the powders. Add Part C to Parts A+B and continue heating. Mix and heat the Part D ingredients to 80°C. Add Part D at 80°C to Part A+B+C also at 80°C. Mix 5 minutes. Blend the Part E ingredients and add them to the batch. Mix for 10 minutes and then begin to cool. When the batch reaches 45°C, Add Parts F and G.

CLEAR SUNSCREEN OIL (SPF ~ 15)

Supplier: Dow Corning Corp.

	Ingredient	Wt.%
A	Cyclomethicone (**Dow Corning 344 Fluid**)	60.00
	Cyclomethicone(and)Dimethiconol (**Dow Corning Q2-1401 Fluid**)	10.00
	Diisopropyl Adipate	10.00
B	Octyl Methoxycinnamate	7.00
	Oxybenzone	3.00
	C12-15 Alcohols Benzoate	10.00
	Fragrance	q.s.
		100.00

Mixing Procedure:

Mix the ingredients in Part A and Part B separately. Add Part B to Part A with agitation.

ANHYDROUS SUNSCREEN GEL

Supplier: Gattefossé s.a.

	Ingredient:	Wt.%
	Petrolatum	15.00
	Tribehenin (**Compritol 888 ATO**)	11.00
	Isostearyl Isostearate	22.00
A	Mineral Oil	43.50
	Octyl Methoxycinnamate	3.00
	Oxybenzone	1.00
	Apricot Kernel Oil(and)Beta-Carotene (**Carotene Huileux 10.000**)	0.50
	Tocopherol Acetate	1.00
B	Borage Seed Oil (**Cevenyl**)	3.00
	Preservative, Fragrance	q.s.
		100.00

Mixing Procedure:

Mix and heat the Part A ingredients to 70°C. Add the **Cevenyl** and preservative with good agitation. Cool while stirring and add the fragrance at 35°C. Homogenize, if necessary.

SUNSCREEN CREAM

Supplier: Gattefossé s.a.

	Ingredient:	Wt.%
A	Propylene Glycol Stearate (**Monosteol**)	5.00
	Glyceryl Stearate (**Geleol**)	5.00
	Octyl Methoxycinnamate	7.00
	Oxybenzone	2.00
	Octyldodecyl Myristate (**M.O.D.**)	5.00
	Dimethicone, 100cs.	2.00
	Apricot Kernel Oil(and)Beta-Carotene (**Carotene Huileux 10.000**)	0.10
B	Demineralized Water	59.70
	Sucroester 7	5.00
	Glycerin	5.00
C	Cyclomethicone(and)Dimethiconol	3.00
	Propylene Glycol(and)Diazolidinyl Urea(and) Methylparaben(and)propylparaben	1.00
	Fragrance	0.20
		100.00

Mixing Procedure:

Mix the Part B ingredients and verify that the **Sucroester 7** is completely dissolved in the aqueous phase. Mix and heat the Part A ingredients to 75°C. Heat the Part B mixture to 75°C. Add Part A to Part B with turbine mixing for 3 minutes. Cool while gently stirring. Add the Part C ingredients at 35°C, in the order shown. Cool to ambient temperature while stirring.

SUNSCREEN LOTION I

Supplier: Gattefossé s.a.

	Ingredient:	Wt.%
A	Glyceryl Stearate(and)Propylene Glycol Stearate(and)Glyceryl Isostearate(and) Propylene Glycol Isostearate(and) Oleth-25(and)Ceteth-25 (**Hydrolactol 70**)	8.00
	Cetyl Alcohol	1.00
	Mineral Oil	6.00
	Mineral Oil(and)Apricot Kernel Oil (and)Calendula Extract (**Vegetol Calendula WL 1072 Oily**)	3.00
	Octyl Methoxycinnamate	7.50
	Oxybenzone	2.00
	Antioxidant	q.s.
B	Demineralized Water	68.15
	Glycerin	3.00
	Tetrasodium EDTA	0.05
	Carbomer 941	0.10
C	Triethanolamine, 50%	0.20
	Borsage Seed Oil (**Cevenyl**)	1.00
	Preservative, Fragrance	q.s.
		100.00

Mixing Procedure:

Disperse the Carbomer 941 in the Water and add the remaining Part B ingredients. Let the mixture stand until entrapped air is released. Heat Parts A and B, separately, to 75°C. With moderate stirring, add Part B to Part A at 75°C. Add the first two Part C ingredients and mix until uniform. Cool while stirring to 30-35°C. Add the remaining Part C ingredients and mix until uniform.

SUNSCREEN LOTION II

Supplier: Gattefossé s.a.

	Ingredient:	Wt.%
A	Glyceryl Stearate(and)Propylene Glycol Stearate(and)Glyceryl Isostearate(and) Propylene Glycol Isostearate(and) Oleth-25(and)Ceteth-25 (**Hydrolactol 70**)	20.00
	Isostearyl Isostearate	8.00
	Titanium Dioxide	25.00
B	Demineralized Water	44.00
	Glycerin	3.00
C	Preservative	q.s.
		100.00

Mixing Procedure:

Heat Parts A and B, separately to 75°C. Add Part B to Part A while stirring. Cool to approximately 30°C and add Part C.

SUNSCREEN OIL

Supplier: Gattefossé s.a

	Ingredient:	Wt.%
A	Coconut Oil	10.00
	Isostearyl Isostearate	20.00
	Oxybenzone	1.00
B	Mineral Oil(and)Apricot Kernel Oil (and)Calendula Extract (**Vegetol Calendula WL 1072 Oily**)	15.00
	Octyl Methoxycinnamate	4.00
	Mineral Oil	50.00
	Antioxidant, Fragrance	q.s.
		100.00

Mixing Procedure:

Mix and heat the Part A ingredients to 40°C. Continue mixing until the Oxybenzone is completely dissolved. Add the Part B ingredients in the order shown and mix until uniform.

MOISTURIZER WITH SUNSCREEN

Supplier: General Electric Company

	Ingredient:	Wt.%
A	Petrolatum	0.88
	Cetyl Alcohol	0.88
	Stearic Acid	0.88
	PEG-8 Stearate	0.88
	Glyceryl Stearate(and)PEG-100 Stearate	0.44
	Dimethicone (**SF-96-100**)	0.29
	C12-15 Alcohols Benzoate	1.54
	Lecithin	0.88
	Corn Oil(and)Glyceryl Oleate(and) Propylene Glycol(and)BHA(and) BHT(and)Propyl Gallate(and)Citric Acid	q.s.
B	Water	82.42
	Carbomer 940	0.23
	Edetate Disodium	0.05
	Methylparaben	0.22
C	Octyl Methoxycinnamate	3.60
	Dimethicone(and)Trimethylsiloxysilicate (**SS4267**)	2.22
	Cyclomethicone (**SF1204**)	4.50
D	Sodium Hydroxide	0.09
		100.00

Mixing Procedure:

Combine and heat the Part A ingredients to 80°C. In a separate vessel, combine and heat the Part B ingredients to 80°C. Add Part A to Part B with mixing. Continue to mix while cooling to 50°C. In a separate vessel, mix the Part C ingredients and add the mixture to the batch. Homogenize in a colloid mill or homogenizer at 5000psi. Cool with agitation to 30°C, add Part D and homogenize. Target pH is 5.3.

PROTECTIVE SKIN PRODUCT WITH SUNSCREEN

Supplier: General Electric Company

	Ingredient:	Wt.%
	Cyclomethicone (**SF1202**)	12.00
	Mineral Oil, Light	1.00
	Octyl Methoxycinnamate	5.00
A	Dimethicone(and)Trimethylsiloxysilicate (**SS4267**)	3.00
	Cyclomethicone(and)Dimethicone Copolyol (**SF1228**)	10.00
	Lanolin	0.50
	Titanium Dioxide, Micronized	3.00
	Polysorbate 80	0.20
	Glycerin	3.00
B	Sodium Chloride	1.00
	Water	61.20
	Quaternium-15	0.10
		100.00

Mixing Procedure:

Mix the Part A ingredients in the order shown, thoroughly mixing each component until homogeneous before adding the next. The solution should be clear until the last ingredient in Part A is added. In a separate vessel combine the Part B ingredients. Add Part B to Part A with good mixing, gradually increasing agitation to high shear as the mixture thickens. Continue agitation for 5-10 minutes. Homogenize for 1-2 minutes.

BROAD-SPECTRUM PROTECTIVE SUN STICK (SPF 11)

Supplier: Givaudan-Roure Corp.

	Ingredient:	Wt.%
	Butyl Methoxydibenzoylmethane (**Parsol 1789**)	2.00
	Octyl Methoxycinnamate (**Parsol MCX**)	7.50
	Castor Oil (**Ricinol**)	7.50
	Hydrogenated Castor Oil	7.50
A	Oleyl Alcohol (**Satol**)	20.00
	Microcrystalline Wax	30.00
	Mineral Oil	7.35
	Petrolatum	8.53
	Dimethicone, 200cs.	3.50
	BHT	0.02
	Beta-Carotene, 1% Solution	0.30
B	**D-Panthenol**	0.30
	Propylene Glycol	3.00
C	Fragrance, Preservatives, Petrolatum to 100%	q.s.
		100.00

Mixing Procedure:

Mix and heat the Part A ingredients to 85°C. Preheat the Part B ingredients, add them to Part A and mix until homogeneous. Add the Part C ingredients and continue mixing. When homogeneous, pour into molds.

BROAD SPECTRUM SUNSCREEN GEL (SPF 18)

Supplier: Givaudan-Roure Corp.

	Ingredient:	Wt.%
A	Alcohol SD	50.00
	Hydroxypropyl Methylcellulose	2.00
	Alcohol SD	27.50
	Butyl Methoxydibenzoylmethane (**Parsol 1789**)	1.00
B	Octyl Methoxycinnamate (**Parsol MCX**)	7.50
	Oxybenzone	4.00
	C12-15 Alcohols Benzoate	5.00
	Phenyl Dimethicone	1.00
C	Perfume, Perfume, Deionized Water to 100.00%	q.s.
		100.00

Mixing Procedure:

Mix the Part A ingredients until homogeneous. Mix the Part B ingredients with mild heating. Add Part B to Part A with mixing. Add Part C and mix until homogeneous.

BROAD-SPECTRUM SUNSCREEN CREAM (SPF ~ 15)

Supplier: Givaudan-Roure Corp.

	Ingredient:	Wt.%
	Butyl Methoxydibenzoylmethane (**Parsol 1789**)	1.50
	Octyl Methoxycinnamate (**Parsol MCX**)	7.50
	Oxybenzone	5.50
A	Dioctyl Maleate	10.00
	PVP/Eicosene Copolymer	2.50
	Dimethicone, 100 cs.	0.50
B	DEA-Cetyl Phosphate (**Amphisol**)	2.50
	Deionized Water	50.00
C	Carbomer 940, 2% Aqueous Dispersion	5.00
	Glycerin	5.10
D	Deionized Water	7.80
	Triethanolamine, 99%	0.10
E	Fragrance, Preservatives, Deionized Water to 100.00%	q.s.
		100.00

Mixing Procedure:

While mixing with a planetary mixer, heat the Part A ingredients to 85°C. Add Part B. When homogeneous, Add the Part C ingredients, previously mixed and heated to 75°C. While mixing, cool to 45°C. Add Parts D and E, compensate for the water loss and continue mixing while cooling to room temperature.

WATERPROOF SUNSCREEN LOTION (SPF 15)

Supplier: BF Goodrich Company

	Ingredient:	Wt.%
	Deionized Water	72.75
A	Hydroxypropyl Methylcellulose, 1% Solution	10.00
	Quaternium-15	0.15
	Edetate Disodium	0.05
	Octyl Methoxycinnamate	7.00
	Octyl Salicylate	3.00
	Oxybenzone	2.00
	C12-15 Alcohols Benzoate	4.00
B	Acrylates/C10-30 Alkyl Acrylates Crosspolymer(**Pemulen TR-2**)	0.25
	Carbomer 954 (**Carbopol 954**)	0.20
	Methylparaben	0.15
	Propylparaben	0.05
C	Triethanolamine, 99%	0.40
		100.00

Mixing Procedure:

Combine the Part A ingredients and mix until homogeneous. Combine the first three Part B ingredients in a separate vessel. Mix until the oxybenzone is dissolved. Warming hastens dissolution. Disperse the last 4 Part B ingredients in the other Part B mixture. With moderate agitation, add Part B to Part A. Mix for 20-40 minutes or until smooth. Add Part C and mix vigorously until a smooth, lustrous product is obtained.

SUNSCREEN CREAM

Supplier: Hüls America, Inc.

	Ingredient:	Wt.%
	Caprylic/Capric Triglyceride(and)Stearalkonium Hectorite(and)Propylene Carbonate (**Myglyol-Gel**)	15.00
A	Isostearyl Diglyceryl Succinate (**Imwitor 780K**)	5.00
	Mineral Oil	5.00
	Isoamyl ρ-Methoxycinnamate	5.00
B	Preservative	q.s.
	Water	70.00
C	Fragrance	q.s.
		100.00

Mixing Procedure:

Mix and heat the Part A ingredients to 75-80°C. Heat the Part B ingredients to the same temperature and gradually add them to Part A with mixing. Cool the batch at 40°C and add Part C.

SUNSCREEN LOTION

Supplier: Hüls America, Inc.

	Ingredient:	Wt.%
	Caprylic/Capric Triglyceride(and)Stearalkonium Hectorite(and)Propylene Carbonate (**Miglyol-Gel**)	4.00
	Caprylic/Capric Triglyceride(**Miglyol 812**)	5.00
	Arlacel 481	3.00
A	PEG-7 Hydrogenated Castor Oil	5.00
	Isopropyl Myristate	12.50
	Paraffin	2.00
	Octyl Methoxycinnamate	7.50
	Butyl Methoxydibenzoylmethane	4.00
	Glycerin	5.00
	Carbomer 934	0.20
B	Magnesium Sulfate	0.70
	Preservative	q.s.
	Water	56.10
C	Fragrance	q.s.
		100.00

Mixing Procedure:

Mix and heat the Part A ingredients to 75-80°C. Mix the Part B ingredients with a high speed stirrer and heat to 75-80°C. Part B is emulsified into Part A. Cool to about 30°C and add Part C.

SUNTAN OIL

Supplier: Inolex Chemical Company

Ingredient:	Wt.%
Jojoba Oil	2.00
Propylene Glycol Dipelargonate(**Lexol PG900**)	5.00
Propylene Glycol Dioctanoate (**Lexol PG800**)	5.00
Propylene Glycol Dicaprylate/Dicaprate (**Lexol PG 865**)	5.00
Homosalate	8.00
Acetylated Lanolin Alcohol	2.00
Mineral Oil	73.00
	100.00

Mixing Procedure:

Blend all the ingredients until a clear solution is obtained. Fill.

SUNTAN LOTION

Supplier: Inolex Chemical Company

	Ingredient:	Wt.%
	Glyceryl Stearate(and)PEG-100 Stearate (**Lexemul 561**)	4.00
	Stearic Acid	2.00
	Cetyl Alcohol	1.00
	Mineral Oil	4.00
A	Sesame Oil	1.00
	Laneth-10 Acetate	0.50
	Ethyl 4-[bis(hydroxypropyl)]Aminobenzoate	3.00
	Mink Oil	0.10
	Polysorbate 20	0.20
	Propylparaben (**Lexgard P**)	0.05
	Propylene Glycol	4.00
	Carbomer 934	0.20
B	Triethanolamine, to pH 7.2-7.5	q.s.
	Aloe Vera Gel	2.00
	Methylparaben (**Lexgrad M**)	0.15
	Water	77.80
		100.00

Mixing Procedure:

Heat the Part B Water to 70°C, disperse the Carbomer 934 in it and mix until dissolved. Neutralize with Triethanolamine. Dissolve the **Lexgard M** in the Propylene Glycol and add to the batch with the remaining Part B ingredients. In a separate vessel, mix and melt the Part A ingredients. Slowly add Part A to Part B with agitation. Cool to the set point and fill.

MOISTURIZING SUN PROTECTION LOTION W/ SEAWEED EXTRACT

Supplier: International Sourcing, Inc.

	Ingredient:	Wt.%
	Mineral Oil	3.50
	Stearic Acid, XXX	1.20
	Cetyl Alcohol	0.40
	Glyceryl Stearate	0.60
A	Dimethicone	2.20
	Methylparaben	0.15
	Propylparaben	0.10
	Octyl Methoxycinnamate	4.00
	Deionized Water	37.50
	Algae Extract(and)Sorbitol (**Omagel SH**)	6.00
B	Propylene Glycol	1.40
	Water(and)Propylene Glycol(and)Aloe Extract(**Dermascreen**)	5.00
	Deionized Water	37.00
C	Carbomer 940	0.40
	Triethanolamine, 98%	0.40
D	Fragrance	0.15
		100.00

Mixing Procedure:

Disperse the Carbomer 940 in water and heat to 80°C. Add Triethanolamine slowly to prepare Part C. Mix the Part B ingredients and heat to 80°C, with agitation, until homogeneous. Add Part B to Part C. Combine the Part A ingredients, heat to 80°C and add them to Parts B+C. Cool with constant stirring and add Part D at about 30°C.

SUNSCREEN MOISTURIZING CREAM

Supplier: International Sourcing, Inc.

	Ingredient:	Wt.%
	Stearic Acid, XXX	2.00
	Glyceryl Stearate	4.20
	Acetylated Lanolin Alcohol	5.00
A	Octyl Methoxycinnamate	6.00
	Oxybenzone	2.00
	Water(and)Propylene Glycol(and)Aloe Extract (**Dermascreen**)	2.50
	Triethanolamine, 98%	0.90
B	Algae Extract(and)Sorbitol (**Omagel SH**)	10.00
	Methylparaben	0.15
	Deionized Water	58.75
C	Propylene Glycol	8.30
	Propylparaben	0.05
D	Fragrance	0.15
		100.00

Mixing Procedure:

Heat the Part B water to 80°C and dissolve the remaining Part B ingredients in it. Mix the Part C ingredients and add the mixture to Part B. Mix and heat the Part A ingredients to 80°C and add the mixture to Parts B+C with constant stirring. Cool to 30-35°C with constant stirring and add Part D.

CLEAR SUNSCREEN GEL (SPF Approx.15)

Supplier: ISP Van Dyk

	Ingredient:	Wt.%
	Carbomer 940	1.50
A	Alcohol SD40	53.50
	Polyoxamine 704	7.00
	Isocetyl Alcohol (**Ceraphyl® ICA**)	10.00
B	Padimate O (**Escalol® 507**)	7.00
	Oxybenzone (**Escalol® 567**)	3.00
C	PEG-15 Cocamine	3.50
D	Deionized Water	14.50
		100.00

Mixing Procedure:

Combine the Part A ingredients at room temperature and mix until smooth. Combine the Part B ingredients at room temperature and add the mixture to Part A while mixing. Add Part C to the mixture slowly, and continue agitation. Add Part D to the mixture very slowly, stirring until uniform. Package.

INEXPENSIVE SUNTAN OIL (SPF Approx. 2)

Supplier: ISP Van Dyk

Ingredient:	Wt.%
C12-15 Alcohols Lactate (**Ceraphyl 41**)	10.00
Octyl Palmitate (**Ceraphyl 368**)	26.75
Mineral Oil, 65/75 cs.	60.00
Padimate O (**Escalol 507**)	3.25
Fragrance	q.s.
	100.00

Mixing Procedure:

Weigh all the ingredients into a suitable vessel and mix until uniform.

SUNSCREEN MOUSSE (SPF Approx. 15)

Supplier: ISP Van Dyk

	Ingredient:	Wt.%
A	Deionized Water	69.30
	Carbomer 934	0.20
	Isostearic Acid	5.00
	Isodecyl Oleate (**Ceraphyl 140A**)	8.00
	Octyl Methoxycinnamate (**Escalol 557**)	7.50
B	Oxybenzone (**Escalol 567**)	3.00
	Propylene Glycol	5.00
	Propylene Glycol(and)Diazolidinyl Urea(and) Methylparaben(and) Propylparaben(**Germaben® II**)	1.00
	Fragrance	q.s.
C	Triethanolamine, 99%	1.00
		100.00

Mixing Procedure:

Disperse the Carbomer 934 in the Water. Add the Part B ingredients in the order listed, mixing after each addition. Slowly add Part C and mix until uniform. Package with 95% concentrate and 5% propellant.

SPF 15+ SUNSCREEN LOTION

Supplier: ISP Van Dyk

	Ingredient:	Wt.%
	Octyl Palmitate (**Ceraphyl 368**)	10.00
	Padimate O (**Escalol 507**)	8.00
	Oxybenzone (**Escalol 567**)	4.00
A	Glyceryl Dilaurate (**Emulsynt® GDL**)	2.00
	Glyceryl Stearate(and)Laureth-23 (**Cerasynt® 945**)	2.00
	Stearic Acid, XXX	1.00
	PEG-40 Stearate	1.00
	Dimethicone Copolyol	1.00
	Deionized Water	56.25
	Carbomer 941	0.25
B	Propylene Glycol	3.00
	Propylparaben	0.20
	Methylparaben	0.20
C	Triethanolamine, 99%	0.60
D	Imidazolidinyl Urea (**Germall®115**)	0.50
	Deionized Water	10.00
		100.00

Mixing Procedure:

Heat the Part A ingredients, with mixing, to 80°C. In a separate vessel, disperse the Carbomer 941 in the Part B Water with vigorous agitation and then heat to 80°C. Add the remaining Part B ingredients and mix until uniform. Slowly Add Part A to Part B with agitation. Add Part C and mix for 30 minutes. Cool to 40°C with agitation, add Part D and continue cooling to 28°C

"PABA-FREE" WATERPROOF SUNSCREEN (SPF Approx. 15)

Supplier: National Starch And Chemical Company

	Ingredient:	Wt.%
	Octyl Methoxycinnamate	7.50
	Oxybenzone	3.00
	Octyl Palmitate	3.00
	Cetyl Alcohol	1.00
A	Stearic Acid, XXX	2.00
	PEG-40 Stearate	1.50
	Dimethicone Copolyol	1.00
	Dimethyl Stearamine	2.00
	Acrylates/t-octylpropenamide Copolymer (**Dermacryl-79**)	2.00
	Deionized Water	74.90
B	Carbomer 941	0.20
	Triethanolamine, 99%	0.70
C	Propylene Glycol(and)Diazolidinyl Urea(and) Methylparaben(and)Propylparaben	1.00
D	Fragrance	0.20
		100.00

Mixing Procedure:

Disperse the Carbomer 941 in the Water and heat to 80°C. Add the Triethanolamine slowly to complete Part B. Combine all the Part A ingredients except **Dermacryl-79** and heat to 80°C. Slowly sift **Dermacryl-79** into the oil phase, with constant stirring, until dissolved. Add Part A to Part B at 80°C and mix for 15 minutes. Cool to 40°C and add Parts C and D with mixing. Cool to room temperature and package.

TITANIUM DIOXIDE-BASED WATERPROOF SUNSCREEN (SPF 12)

Supplier: National Starch And Chemical Company

	Ingredient:	Wt.%
	Isocetyl Alcohol	7.00
	C12-15 Alcohols Benzoate	2.00
	Stearic Acid	2.00
	PEG-40 Stearate	2.00
	Dimethicone Copolyol	1.00
A	Cetyl Alcohol	1.00
	Glyceryl Stearate	0.50
	Dimethyl Stearamine	2.00
	Acrylates/Octylacrylamide Copolymer (**Dermacryl-79**)	2.00
	Titanium Dioxide(and) C12-15 Alcohols Benzoate	10.00
	Deionized Water	59.30
	Carbomer 941, 2% Aqueous Solution	10.00
B	Methylparaben	0.15
	Propylparaben	0.10
	Triethanolamine, 99%	0.80
C	Diazolidinyl Urea	0.15
D	Fragrance	q.s.
		100.00

Mixing Procedure:

Combine the Part B ingredients and heat to 80°C. In a separate vessel, combine all the Part A ingredients, except **Dermacryl-79** and the Titanium Dioxide-containing component, and heat them to 80°C. Slowly sift **Dermacryl-79** into the mixture and stir until completely dissolved. Add the Titanium Dioxide-containing component with constant stirring until completely dispersed. Add Part A to Part B at 80°C and mix for 30 minutes. Cool to 40°C and add Parts C and D. Homogenize, cool to room temperature and package.

WATERPROOF SUNSCREEN SPRAY MIST (SPF 16)

Supplier: National Starch And Chemical Company

	Ingredient:	Wt.%
A	Alcohol SD40	64.20
	Deionized Water	5.00
	Triethanolamine, 99%	0.30
	Acrylates/Octylacrylamide Copolymer (**Dermacryl-79**)	1.00
B	Octyl Methoxycinnamate	7.50
	Oxybenzone	3.00
	Menthyl Anthranilate	3.50
	Cyclomethicone	5.00
	C12-15 Alcohols Benzoate	10.00
	Tocopherol Acetate	0.50
C	Fragrance	q.s.
		100.00

Mixing Procedure:

Combine the first three Part A ingredients. Slowly sift in **Dermacryl-79** and mix until completely dissolved. Mix and add the Part B ingredients. Mix until uniform. Add Part C, mix until uniform and package.

HYDROALCOHOLIC SUNTAN LOTION

Supplier: NIPA Laboratories, Inc.

	Ingredient:	Wt.%
A	Panthenol	0.50
B	Octyl Methoxycinnamate	1.00
C	Diethanolamine Methoxycinnamate (**Nipasorb D**)	2.00
D	Ricinoleyl Alcohol	0.20
E	Alcohol, 95%	60.00
F	Fragrance, Solubilizer, Preservative	q.s.
G	Distilled Water	36.30
		100.00

Mixing Procedure:

Mix Parts A through F until all the ingredients are dissolved. Add Part G slowly with mixing. Store in a cool location, then filter.

SUNSCREEN OIL

Supplier: R.I.T.A Corp.

	Ingredient:	Wt.%
A	Isopropyl Myristate	10.00
B	Lanolin Oil (**Ritalan**)	3.00
C	Acetylated Lanolin (**Ritacetyl**)	1.75
D	Polysorbate 80(and)Acetylated Lanolin Alcohol(and)Cetyl Acetate (**Ritawax AEO**)	5.00
E	Mineral Oil	72.15
F	Propylparaben	0.10
G	Fragrance	q.s.
H	Homosalate	8.00
		100.00

Mixing Procedure:

Mix Parts A through F, stir constantly and begin heating to 67-72°C. Hold at that temperature for 20 minutes, stirring until all the ingredients dissolve. Begin cooling, and add Parts G and H at 40-43°C. Cool to 25-30°C and package.

WATERPROOF SUNSCREENING CREAM (O/W) SPF 8

Supplier: R.I.T.A Corp.

	Ingredient:	Wt.%
	Padimate O	8.00
	Stearic Acid	4.00
	Cetyl Alcohol	1.00
	Pationic ISL	3.00
	Pationic SSL	3.00
	PVP/Eicosene Copolymer	5.00
	Dimethicone	0.30
	Shea Butter (**Shebu Refined**)	3.00
B	Cetearyl Alcohol(and)Ceteareth-20 (**Ritapro 300**)	2.50
	Distilled Water	64.00
C	Glycerin	5.00
	Carbomer 940 (**Acritamer 940**)	0.10
D	Distilled Water	1.00
	Triethanolamine, 99%	0.10
E	Fragrance, Preservatives	q.s.
		100.00

Mixing Procedure:

Heat the Part A ingredients to 85°C while mixing with a planetary mixer. Add Part B. Mix the Part C ingredients and heat to 75°C. When homogeneous, add Parts A+B to Part C with mixing. Mix until uniform. Mix and add the Part D ingredients to A+B+C. Begin cooling and add Part E at 40°C. Compensate for water loss and continue mixing while cooling to 35°C.

SUNSCREEN - HIGH SPF

Supplier: Rohm and Haas Company

	Ingredient:	Wt.%
	Deionized Water	61.32
A	**Aculyn 33**	3.33
	Propylene Glycol	2.00
	Tetrasodium EDTA	0.10
	Octyl Methoxycinnamate	7.50
	Oxybenzone	6.00
	Octocrylene	8.00
B	C12-15 Alcohols Benzoate	2.00
	PVP/Eicosene Copolymer	3.00
	Cyclomethicone	2.00
	PEG-40 Stearate	1.50
	Cetearyl Alcohol(and)Ceteareth-20	1.50
C	Triethanolamine, 99%	0.75
D	Preservative	1.00
		100.00

Mixing Procedure:

Mix and heat the Part A ingredients to 75°C. Mix and heat the Part B ingredients to 75°C. Add Part B to Part A. Add Part C and cool to 45°C. Add Part D.

PABA-FREE SUNBLOCK/EMOLLIENT (SPF Approx. 15)

Supplier: Sandoz Chemicals Corp.

	Ingredient:	Wt.%
A	Isopropyl PPG-2 Isodeceth-7 Carboxylate (**Velsan D8P3**)	5.00
	Oxybenzone	3.00
	Cetyl PPG-2 Isodeceth-7 Carboxylate (**Velsan D8P16**)	0.50
	Octyldodecanol	2.50
B	Octyl Methoxycinnamate	7.00
	Dimethicone	1.00
	Stearyl Alcohol	2.50
	Stearic Acid, XXX	2.20
	Deionized Water	61.88
	Magnesium Aluminum Silicate	1.25
	Cellulose Gum, High Viscosity	0.12
C	PVP/Eicosene Copolymer	3.00
	Triethanolamine, 99%	0.40
	Edetate Disodium	0.10
	PEG-8	4.00
	Propylene Glycol	5.00
D	Imidazolidinyl Urea	0.50
	Fragrance	0.05
		100.00

Mixing Procedure:

Premix the Part A ingredients. Add the Part B ingredients. Prepare Part C by hydrating the Magnesium Aluminum Silicate and Cellulose Gum in water using a blender or "Jiffy" mixer. Add the remaining Part C ingredients. Heat Parts A+B and Part C, separately, to 77°C with stirring. Add Parts A+B to Part C slowly, then discontinue heating. Continue stirring A+B+C until the batch cools to 50°C. Add the Part D ingredients. Mix while cooling to 35-40°C and package.

SUNSCREEN CREAM

Supplier: Southern Clay Products Company

	Ingredient:	Wt.%
	Montmorillinite (**Gelwhite GP**)	2.00
A	Water	67.00
	Propylene Glycol	3.00
	Triethanolamine	1.00
	Mineral Oil, Light	5.00
	Isopropyl Myristate	5.00
B	Acetylated Lanolin Alcohol	5.00
	Stearic Acid	3.00
	Cetyl Alcohol	2.00
	Octyl Methoxycinnamate	7.00
C	Preservatives	q.s.
		100.00

Mixing Procedure:

Slowly add the **Gelwhite GP** to the Water, mixing with maximum available shear. Add the remaining Part A ingredients in the order shown, mixing each until smooth. Heat Part A to 75°C. Combine the Part B ingredients and heat them to 75°C. Add Part B to Part A, mixing until smooth and uniform. Continue mixing while cooling to 50°C then reduce mixer speed and stir until cool. Add Part C with slow mixing until smooth and uniform.

WATER-RESISTANT SUNTAN LOTION (SPF Approx. 5)

Supplier: R.T. Vanderbilt Company, Inc.

	Ingredient:	Wt.%
	Magnesium Aluminum Silicate (**Veegum**)	1.75
A	Glycerin	2.25
	Deionized Water	75.50
	Ethyl 4-[Bis(Hydroxypropyl)]Aminobenzoate	2.00
	Sorbitan Laurate	3.50
B	Polysorbate 20	4.50
	Phenyl Trimethicone	5.00
	Stearic Acid, XXX	5.50
C	Preservative	q.s.
		100.00

Mixing Procedure:

Slowly add the **Veegum** to the water while agitating with maximum available shear. Continue mixing until smooth. Add the remaining Part A ingredient and mix until uniform. Heat Part A to 70°C. Mix and heat Part B ingredients to 75°C. Add Part B to Part A, mixing until cool. Add Part C and mix until uniform.

SUNSCREEN CREAM (SPF Approx. 12)

Supplier: R.T. Vanderbilt Company, Inc.

	Ingredient:	Wt.%
	Tromethamine Magnesium Aluminum Silicate (**Veegum PRO**)	1.50
A	Water	67.70
	Propylene Glycol	3.00
	Triethanolamine	0.60
	Oxybenzone	5.00
	Octyl Methoxycinnamate	7.50
	C12-15 Alcohols Benzoate	7.50
B	Mineral Oil(and)Lanolin Alcohol	4.00
	Stearic Acid, XXX	2.00
	C18-36 Acid	0.20
	Glycol Stearate SE	0.50
	Cetyl Alcohol	0.50
C	Preservative, Dye, Fragrance	q.s.
		100.00

Mixing Procedure:

Heat the Part A Water to 75-80°C. Slowly add the **Veegum PRO** while agitating at maximum available shear. Mix until smooth. Add the remaining Part A ingredients in the order shown with careful mixing of each until smooth. Maintain temperature at 75-80°C. Combine the Part B ingredients and heat to 75-80°C. Add Part B to Part A and mix while cooling. When cool, add the Part C ingredients.

ULTRA **VIOLET ABSORBING SUNSCREEN (SPF Approx. 15)**

Supplier: R.T. Vanderbilt Company, Inc.

	Ingredient:	Wt.%
	Magnesium Aluminum Silicate (**Veegum *Ultra***)	1.50
A	Deionized Water	70.50
	Glycerin	5.50
	PEG-150 Distearate	3.00
	Dioctyl Malate	2.00
	Mineral Oil	4.00
B	Cetyl Alcohol	0.50
	Oxybenzone	3.00
	Padimate O	7.00
	Steareth-2	0.90
	Steareth-20	2.10
C	Preservative, Fragrance	q.s.
		100.00

Mixing Procedure:

Heat the Part A Water to 55°C. Slowly add **Veegum *Ultra*** to the Water while stirring with a propeller mixer at 500rpm. Increase the mixer speed to 1500-1700rpm. and mix for 30 minutes maintaining the temperature at 55°C. Add the remaining Part A ingredient and mix 5 minutes. Mix the Part B ingredients and heat them to 60°C. Add Part B to Part A while mixing at 1500-1700rpm. Mix for 30 minutes. Avoid air entrapment. Slow the mixer to 1000rpm and continue mixing while cooling to 35°C. Add Part C and mix until uniform. Package.

[1] *Federal Register*, 50, No. 90, May 12, 1993, pp 28295-28302

CHAPTER 20

VITAMINS AND MINERALS

Although there are several of brand name products that contain only one vitamin (i.e. Vitamin C) or one mineral (i.e. Calcium), by far the most common types of products are multivitamins and multivitamins with minerals. About 90% of the products in this chapter fall into these two categories. Some products contain as many as 32 vitamins and mineral ingredients

Among the vitamin ingredients usually found in the multivitamin products are; Vitamin A (as the Acetate or Beta-Carotene), Vitamin B_1 (as Thiamine Mononitrate), Vitamin B_2 (Riboflavin), Vitamin B_6 (as Pyridoxine Hydrochloride), Vitamin B_{12} (Cyanocobalamin), Niacin (as Niacinamide), Pantothenic Acid and Folic Acid. These last three are part of The B Complex group. Other vitamins frequently incorporated are Vitamin C (Ascorbic Acid), Vitamin D (Ergocalciferol or Cholecalciferol), Vitamin E (Alpha-Tocopheryl Acetate), Vitamin K_1 (as Phytonadione) and Biotin (a.k.a. Vitamin H),

Mineral ingredients are usually added as the organic or inorganic salt or the oxide of the element.

The label of most products contains a table of the active ingredients as well as a listing of all the ingredients in the product. The concentration/unit dose of the active ingredients is expressed in the International Unit of potency, abbreviated "I.U." or in milligrams (10^{-3} grams), abbreviated "mg" or in micrograms (10^{-6} grams), abbreviated "mcg".

Product forms are usually chewable tablets, caplets or gelcaps although products designed primarily for infants and young children are frequently supplied as liquids, to facilitate ingestion.

PART I

BRAND NAME FORMULATIONS

ALLBEE C-800

Manufacturer: A.H. Robins Company

Product Form: Tablet

Active Ingredient:	Concentration: (Per Tablet)
Vitamin E	45 I.U.
Vitamin C	800mg
Vitamin B_1 (Thiamine)	15mg
Vitamin B_2 (Riboflavin)	17mg
Niacin	100mg
Vitamin B_6	25mg
Vitamin B_{12}	12mcg
Pantothenic Acid	25mg

All Ingredients:

Ascorbic Acid, Niacinamide Ascorbate, Modified Starch, Vitamin E Acetate, Hydrolyzed Protein, Calcium Pantothenate, Hydroxypropyl Methylcellulose, Pyridoxine Hydrochloride, Riboflavin, Stearic Acid, Thiamine Mononitrate, Color, Silicon Dioxide, Lactose, Magnesium Stearate, Povidone, Polyethylene Glycol 400 or 4000, Vanillin, Gelatin, Polysorbate 20 or 80, Sorbic Acid, Sodium Benzoate, Cyanocobalamin. May also contain Hydroxypropylcellulose and Propylene Glycol.

ALLBEE C-800 WITH IRON

Manufacturer:	Product Form:
A.H. Robins Company	Tablet

Active Ingredient:	Concentration: (Per Tablet)
Vitamin E	45 I.U.
Vitamin C	800mg
Folic Acid	0.4mg
Vitamin B_1 (Thiamine)	15mg
Vitamin B_2 (Riboflavin)	17mg
Niacin	100mg
Vitamin B_6	25mg
Vitamin B_{12}	12mcg
Pantothenic Acid	25mg
Iron	27mg

All Ingredients:

Ascorbic Acid, Niacinamide Ascorbate, Ferrous Fumarate, Modified Starch, Vitamin E Acetate, Hydrolyzed Protein, Calcium Pantothenate, Hydroxypropyl Methylcellulose, Pyridoxine Hydrochloride, Riboflavin, Stearic Acid, Thiamine Mononitrate, Color, Silicon Dioxide, Lactose, Magnesium Stearate, Povidone, Polyethylene Glycol 400 or 4000, Vanillin, Gelatin, Polysorbate 20 or 80, Sorbic Acid, Sodium Benzoate, Cyanocobalamin. May also contain Hydroxypropylcellulose and Propylene Glycol.

ALLBEE WITH C

Manufacturer:	Product Form:
A.H. Robins Company	Caplet

Active Ingredient:	Concentration: (Per Caplet)
Vitamin C	300mg
Vitamin B_1 (Thiamine)	15mg
Vitamin B_2 (Riboflavin)	10.2mg
Niacin	50mg
Vitamin B_6	5mg
Pantothenic Acid	10mg

All Ingredients:

Niacinamide Ascorbate, Ascorbic Acid, Microcrystalline Cellulose, Corn Starch, Thiamine Mononitrate, Calcium Pantothenate, Riboflavin, Hydroxypropyl Methylcellulose, Pyridoxine Hydrochloride, Magnesium Stearate, Silicon Dioxide, Propylene Glycol, Lactose, Methacrylic Acid, Triethyl Citrate, Titanium Dioxide, Polysorbate 20, Flavor, Saccharin Sodium, Sodium Sorbate.

BEELITH

Manufacturer:	Product Form:
Beach Pharmaceuticals	Tablet

Active Ingredient:	Concentration: (Per Tablet)
Pyridoxine Hydrochloride	25mg
Magnesium Oxide	600mg

Other Ingredients:

Castor Oil, Hydroxypropyl Methylcellulose, Magnesium Stearate, Microcrystalline Cellulose, Pharmaceutical Glaze, Povidone, Sodium Starch Glycolate, D&C Yellow No. 10, FD&C Yellow No. 6, Titanium Dioxide.

BUGS BUNNY & FLINTSTONES, PLUS IRON

Manufacturer:
Miles, Inc.

Product Form:
Tablet, Ch.

Ingredient:	Concentration: (Per Tablet)
Vitamin A (as Acetate and Beta Carotene)	2500 I.U.
Vitamin D	400 I.U.
Vitamin E	15 I.U.
Vitamin C	60mg
Folic Acid	0.3mg
Thiamine	1.05mg
Riboflavin	1.20mg
Niacin	13.50mg
Vitamin B_6	1.05mg
Vitamin B_{12}	4.5mcg
Iron (Elemental)	15mg

All Ingredients:
Sucrose, Sodium Ascorbate, Ferrous Fumarate, Stearic Acid, Gelatin, Flavor, Maltodextrins, Vitamin E Acetate, Niacinamide, Magnesium Stearate, Colors, Glycerides of Stearic and Palmitic Acid, Pyridoxine Hydrochloride, Riboflavin, Thiamine Mononitrate, Vitamin A Acetate, Beta-Carotene, Folic Acid, Vitamin D, Vitamin B_{12}.

BUGS BUNNY & FLINTSTONES, WITH EXTRA C

Manufacturer:
Miles, Inc.

Product Form:
Tablet, Ch.

Ingredient:	Concentration: (Per Tablet)
Vitamin A (as Acetate and Beta Carotene)	2500 I.U.
Vitamin D	400 I.U.
Vitamin E	15 I.U.
Vitamin C	250mg
Folic Acid	0.3mg
Thiamine	1.05mg
Riboflavin	1.20mg
Niacin	13.50mg
Vitamin B_6	1.05mg
Vitamin B_{12}	4.5mcg

All Ingredients:
Sorbitol, Sodium Ascorbate, Ascorbic Acid, Starch, Stearic Acid, Flavor, Gelatin, Vitamin E Acetate, Magnesium Stearate, Niacinamide, Colors, Glycerides of Stearic and Palmitic Acids. Xylitol, Aspartame, Citric Acid, Silica, Pyridoxine Hydrochloride, Vitamin A Acetate, Riboflavin, Malic Acid, Thiamine Mononitrate, Monoammonium Glycyrrhizinate, Folic Acid, Beta-Carotene, Vitamin D, Vitamin B_{12}. Contains Phenylalanine.

BUGS BUNNY & FLINTSTONES, COMPLETE

Manufacturer:
Miles, Inc.

Product Form:
Tablet, Ch.

Ingredient:	Concentration: (Per Tablet)
Vitamin A (as Acetate and Beta Carotene)	5000 I.U.
Vitamin D	400 I.U.
Vitamin E	30 I.U.
Vitamin C	60mg
Folic Acid	0.4mg
Vitamin B_1 (Thiamine)	1.5mg
Vitamin B_2 (Riboflavin)	1.7mg
Niacin	20mg
Vitamin B_6 (Pyridoxine)	2mg
Vitamin B_{12} (Cyanocobalamin)	6mcg
Biotin	40mcg
Pantothenic Acid	10mg
Iron (Elemental)	18mg
Calcium	100mg
Copper	2mg
Phosphorous	100mg
Iodine	150mcg
Magnesium	20mg
Zinc	15mg

All Ingredients:
Dicalcium Phosphate, Sorbitol, Magnesium Phosphate, Sodium Ascorbate, Gelatin, Ferrous Fumarate, Stearic Acid, Starch, Vitamin E Acetate, Citric Acid, Carrageenan, Niacinamide, Magnesium Stearate, Hydrogenated Vegetable Oil, Zinc Oxide, Flavors, Calcium Pantothenate, Colors, Glycerides of Stearic and Palmitic Acids, Silica, Xylitol, Aspartame, Cupric Oxide, Pyridoxine Hydrochloride, Riboflavin, Thiamine Mononitrate, Beta-Carotene, Vitamin A Acetate, Monoammonium Glycyrrhizinate, Folic Acid. Potassium Iodide, Biotin, Vitamin D, Magnesium Oxide, Vitamin B_{12}. Contains Phenylalanine.

CALTRATE 600

Manufacturer:	Product Form:
Lederle Laboratories	Tablet

Ingredient:	Concentration: (Per Tablet)
Calcium Carbonate	1500mg

Other Ingredients:
Croscarmellose Sodium, Hydroxypropyl Methylcellulose, Magnesium Stearate, Microcrystalline Cellulose, PVPP, Sodium Lauryl Sulfate, Titanium Dioxide.

CALTRATE 600 + IRON & VITAMIN D

Manufacturer:	Product Form:
Lederle Laboratories	Tablet

Ingredient:	Concentration: (Per Tablet)
Calcium Carbonate	1500mg
Iron (Elemental, in Time Release Form)	18mg
Vitamin D	125 I.U.

Other Ingredients:
Colors, Croscarmellose Sodium, Hydroxypropyl Methylcellulose, Magnesium Stearate, Microcrystalline Cellulose, Polysorbate 80, Povidone, PVPP, Sodium Lauryl Sulfate, Titanium Dioxide, Triethyl Citrate.

CALTRATE 600 + VITAMIN D

Manufacturer:	Product Form:
Lederle Laboratories	Tablet

Ingredient:	Concentration: (Per Tablet)
Calcium Carbonate	1500mg
Vitamin D	125 I.U.

Other Ingredients:
Colors, Croscarmellose Sodium, Hydroxypropyl Methylcellulose, Magnesium Stearate, Microcrystalline Cellulose, Povidone, PVPP, Sodium Lauryl Sulfate, Titanium Dioxide.

CENTRUM

Manufacturer:
Lederle Laboratories

Product Form:
Tablet

Ingredient:	Concentration: (per Tablet)
Vitamin A (as Acetate and Beta-Carotene)	5000 I.U.
Vitamin D	400 I.U.
Vitamin K_1	25mcg
Vitamin E (as *dl*-Alpha Tocopheryl Acetate)	30 I.U.
Vitamin C	60mg
Folic Acid	400mcg
Vitamin B_1	1.5mg
Vitamin B_2	1.7mg
Niacinamide	20mg
Vitamin B_6	2mg
Vitamin B_{12}	6mcg
Pantothenic Acid	10mg
Biotin	30mcg
Calcium	162mg
Phosphorous	109mg
Iodine	150mcg
Iron	18mg
Magnesium	100mg
Copper	2mg
Zinc	15mg
Manganese	2.5mg
Potassium	40mg
Chloride	36.3mg
Chromium	25mcg
Molybdenum	25mcg
Selenium	20mcg
Nickel	5mcg
Tin (as Stannous Chloride)	10mcg
Silicon	2mg
Vanadium	10mcg
Boron	150mcg

Other Ingredients:
FD&C Yellow No. 6, Hydroxypropyl Methylcellulose, Lactose, Magnesium Stearate, Microcrystalline Cellulose, Polysorbate 80, Povidone, Stearic Acid, Titanium Dioxide, Triethyl Citrate.

CENTRUM, LIQUID

Manufacturer: Lederle Laboratories

Product Form: Liquid

Ingredient:	Concentration: (Per 15ml)
Vitamin A (as Palmitate)	2500 I.U.
Vitamin E (as *dl*-Alpha Tocopheryl Acetate)	30 I.U.
Vitamin C (as Ascorbic Acid)	60mg
Vitamin B_1 (as Thiamine Hydrochloride)	1.5mg
Vitamin B_2 (as Riboflavin)	1.7mg
Niacinamide	20mg
Vitamin B_6 (as Pyridoxine Hydrochloride)	2mg
Vitamin B_{12} (as Cyanocobalamin)	6mcg
Vitamin D_2	400 I.U.
Biotin	300mcg
Pantothenic Acid (as Panthenol)	10mg
Iodine (as Potassium Iodide)	150mcg
Iron (as Ferrous Gluconate)	9mg
Zinc (as Zinc Gluconate)	3mg
Manganese (as Manganese Chloride)	2.5mg
Chromium (as Chromium Chloride)	25mcg
Molybdenum (as Sodium Molybdate)	25mcg

Other Ingredients:

Alcohol(6.6%), Flavors, Citric Acid, Glycerin, Polysorbate 80, Sodium Benzoate, Sucrose.

CENTRUM, JR. + EXTRA C

Manufacturer: Lederle Laboratories

Product Form: Tablet, Ch.

Ingredient:	Concentration: (Per Tablet)
Vitamin A (as Acetate)	5000 I.U.
Vitamin E (as Acetate)	30 I.U.
Vitamin D	400 I.U.
Vitamin C (as Ascorbic Acid and Sodium Ascorbate)	300mg
Folic Acid	400mcg
Thiamine (as Thiamine Mononitrate)	1.5mg
Riboflavin	1.7mg
Niacinamide	20mg
Vitamin B_6	2mg
Vitamin B_{12} (as Cyanocobalamin)	6mcg
Vitamin K_1 (as Phytonadione)	10mcg
Biotin	45mcg
Pantothenic Acid (as Calcium Pantothenate)	10mg
Iodine (as Potassium Iodide)	150mcg
Iron (as Ferrous Fumarate)	18mg
Magnesium (as Magnesium Oxide)	40mg
Copper (as Cupric Oxide)	2mg
Phosphorous (as Tribasic Calcium Phosphate)	50mg
Calcium (as Tribasic Calcium Phosphate)	108mg
Zinc (as Zinc Oxide)	15mg
Manganese (as Manganese Sulfate)	1mg
Chromium (as Chromium Chloride)	20mcg
Molybdenum (as Sodium Molybdate)	20mcg

Other Ingredients:
Flavors, Aspartame, Colors, Citric Acid, Lactose, Magnesium Stearate, Microcrystalline Cellulose, Pregelatinized Starch, Silica, Sorbitol, Stearic Acid, Sucrose.

CENTRUM, JR. + EXTRA CALCIUM

Manufacturer: Lederle Laboratories

Product Form: Tablet, Ch.

Ingredient:	Concentration: (Per Tablet)
Vitamin A (as Acetate)	5000 I.U.
Vitamin E (as Acetate)	30 I.U.
Vitamin D	400 I.U.
Vitamin C (as Ascorbic Acid)	60mg
Folic Acid	400mcg
Thiamine (as Thiamine Mononitrate)	1.5mg
Riboflavin	1.7mg
Niacinamide	20mg
Vitamin B_6	2mg
Vitamin B_{12} (as Cyanocobalamin)	6mcg
Vitamin K_1 (as Phytonadione)	10mcg
Biotin	45mcg
Pantothenic Acid (as Calcium Pantothenate)	10mg
Iodine (as Potassium Iodide)	150mcg
Iron (as Ferrous Fumarate)	18mg
Magnesium (as Magnesium Oxide)	40mg
Copper (as Cupric Oxide)	2mg
Phosphorous (as Dibasic Calcium Phosphate)	50mg
Calcium (as Dibasic Calcium Phosphate and Calcium Carbonate)	160mg
Zinc (as Zinc Oxide)	15mg
Manganese (as Manganese Sulfate)	1mg
Chromium (as Chromium Chloride)	20mcg
Molybdenum (as Sodium Molybdate)	20mcg

Other Ingredients:
Flavors, Citric Acid, Lactose, Magnesium Stearate, Microcrystalline Cellulose, Modified Food Starch, Color, Silica Gel, Sodium Starch Glycolate, Stearic Acid, Sucrose.

CENTRUM, JR. + IRON

Manufacturer:	Product Form:
Lederle Laboratories	Tablet, Ch.

Active Ingredient:	Concentration: (Per Tablet)
Vitamin A (as Acetate)	5000 I.U.
Vitamin E (as Acetate)	30 I.U.
Vitamin D	400 I.U.
Vitamin C (as Ascorbic Acid)	60mg
Folic Acid	400mcg
Thiamine (as Thiamine Mononitrate)	1.5mg
Riboflavin	1.7mg
Niacinamide	20mg
Vitamin B_6	2mg
Vitamin B_{12} (as Cyanocobalamin)	6mcg
Vitamin K_1 (as Phytonadione)	10mcg
Biotin	45mcg
Pantothenic Acid (as Calcium Pantothenate)	10mg
Iodine (as Potassium Iodide)	150mcg
Iron (as Ferrous Fumarate)	18mg
Magnesium (as Magnesium Oxide)	40mg
Copper (as Cupric Oxide)	2mg
Phosphorous (as Dibasic Calcium Phosphate)	50mg
Calcium (as Dibasic Calcium Phosphate and Calcium Carbonate)	108mg
Zinc (as Zinc Oxide)	15mg
Manganese (as Manganese Sulfate)	1mg
Chromium (as Chromium Chloride)	20mcg
Molybdenum (as Sodium Molybdate)	20mcg

Other Ingredients:

Acacia, Flavors, Colors, Colloidal Silicon Dioxide, Dextrins, Dextrose, Gelatin, Hydrogenated Vegetable Oil, Hydrolyzed Protein, Lactose, Magnesium Stearate, Methylparaben, Microcrystalline Cellulose, Modified Food Starch, Mono- and Di-glycerides, Potassium Sorbate, Propylparaben, Sodium Benzoate, Sorbic Acid, Stearic Acid, Sucrose, Sodium Starch Glycolate.

CENTRUM SILVER

Manufacturer:
Lederle Laboratories

Product Form:
Tablet

Active Ingredient	Concentration (Per Tablet)	Active Ingredient	Concentration (Per Tablet)
Vitamin A (Acetate & Beta-Carotene	6000 I.U.	Copper	2mg
Vitamin B_1	1.5mg	Iodine	150mcg
Vitamin B_2	1.7mg	Iron	9mg
Vitamin B_6	3mg	Magnesium	100mg
Vitamin B_{12}	25mcg	Phosphorous	48mg
Boitin	30mcg	Zinc	15mg
Folic Acid	200mcg	Chloride	72mg
Niacinamide	20mg	Chromium	100mcg
Pantothenic Acid	10mg	Manganese	2.5mg
Vitamin C	60mg	Molybdenum	25mcg
Vitamin D	400 I.U.	Nickel	5mcg
Vitamin E	45 I.U.	Potassium	80mg
Vitamin K_1	10mcg	Selenium	25mcg
Calcium	200mg	Silicon	10mcg
Boron	150mg	Vanadium	10mcg

Other Ingredients:
Colors, Crospovidone, Hydroxypropyl Methylcellulose, Lactose, Magnesium Stearate, Microcrystalline Cellulose, Polyethylene Glycol, Polysorbate 80, Silica Gel, Stearic Acid, Titanium Dioxide.

FEMIRON

Manufacturer:
Menley & James Laboratories, Inc.

Product Form:
Tablet

Ingredient	Concentration (Per Tablet)
Vitamin A	5000 I.U.
Vitamin D	400 I.U.
Vitamin E	15 I.U.
Vitamin C	60mg
Folic Acid	0.4mg
Vitamin B_1	1.5mg
Vitamin B_2	1.7mg
Niacinamide	20mg
Vitamin B_6	2mg
Vitamin B_{12}	6mcg
Pantothenic Acid	10mg
Iron	20mg

All Ingredients:
Calcium Carbonate Ferrous Fumarate, Ascorbic Acid Gelatin, Niacinamide, Vitamin E Acetate, Starch, Calcium Pantothenate, Microcrystalline Cellulose, Calcium Silicate, Alginic Acid, Hydroxypropyl Methylcellulose, Crospovidone, Stearic Acid, Sodium Lauryl Sulfate, Vitamin A Acetate, Pyridoxine Hydrochloride, Colors, Riboflavin, Thiamine Mononitrate, Polyethylene Glycol, Magnesium Stearate, Folic Acid, Polysorbate 80. Vitamin D, Cyancobalamin.

FILIBON

Manufacturer: Lederle Laboratories

Product Form: Tablet

Ingredient:	Concentration: (Per Tablet)
Vitamin A (as Acetate)	5000 I.U.
Vitamin D_2	400 I.U.
Vitamin E (as *dl*-Alpha Tocopheryl Acetate	30 I.U.
Vitamin C (as Ascorbic Acid)	60mg
Folic Acid	0.4mg
Vitamin B_1 (as Thiamine Mononitrate)	1.5mg
Vitamin B_2 (as Riboflavin)	1.7mg
Niacinamide	20mg
Vitamin B_6	2mg
Vitamin B_{12} (as Cyanocobalamin)	6mcg
Calcium (as Calcium Carbonate)	125mg
Iodine (as Potassium Iodide)	150mcg
Iron (as Ferrous Fumarate)	18mg
Magnesium (as Magnesium Oxide)	100mg

Other Ingredients:
Ethylcellulose, Hydroxypropyl Methylcellulose, Lactose, Magnesium Stearate, Microcrystalline Cellulose, Povidone, Pregelatinized Starch, Color, Silicon Dioxide, Sodium Lauryl Sulfate, Sodium Starch Glycolate, Titanium Dioxide, Stearic Acid.

GARFIELD

Manufacturer: Menley & James Laboratories, Inc.

Product Form: Tablet, Ch.

Ingredient	Concentration (Per Tablet)	Ingredient	Concentration (Per Tablet)
Vitamin A (as Acetate)	1250 I.U.	Vitamin B_1 (as Thiamine)	1.05mg
Vitamin A (as Beta-Carotene)	1250 I.U.	Vitamin B_2 (Riboflavin)	1.2mg
Vitamin D	400 I.U.	Niacin	13.5mg
Vitamin E	15 I.U.	Vitamin B_6	1.05mg
Vitamin C	60mg	Vitamin B_{12}	4.5mcg
Folic Acid	0.3mg		

All Ingredients:

Sucrose, Ascorbic Acid, Mono- and Di-glycerides, Talc, Gelatin, Vitamin E Acetate, Niacinamide, Flavors, Starch, Calcium Carbonate, Colors, Silica, Stearic Acid, Magnesium Stearate, Pyridoxine Hydrochloride, Riboflavin, Thiamine Mononitrate, Beta-Carotene, Lactose, Vitamin A Acetate, Folic Acid, Vitamin D, Vitamin B_{12}.

GARFIELD, PLUS EXTRA C

Manufacturer: Menley & James Laboratories, Inc.

Product Form: Tablet, Ch.

Ingredient	Concentration (Per Tablet)	Ingredient	Concentration (Per Tablet)
Vitamin A (as Acetate)	1250 I.U.	Vitamin B_1 (as Thiamine)	1.05mg
Vitamin A (as Beta-Carotene)	1250 I.U.	Vitamin B_2 (Riboflavin)	1.2mg
Vitamin D	400 I.U.	Niacin	13.5mg
Vitamin E	15 I.U.	Vitamin B_6	1.05mg
Vitamin C	250mg	Vitamin B_{12}	4.5mcg
Folic Acid	0.3mg		

All Ingredients:

Sucrose, Sodium Ascorbate, Ascorbic Acid, Mono- and Di-glycerides, Microcrystalline Cellulose Talc, Gelatin, Vitamin E Acetate, Niacinamide, Flavors, Starch, Colors, Silica, Magnesium Stearate, Hydrogenated Cotton Seed Oil, Pyridoxine Hydrochloride, Riboflavin, Thiamine Mononitrate, Beta-Carotene, Lactose, Vitamin A Acetate, Folic Acid, Vitamin D, Vitamin B_{12}.

GARFIELD, PLUS IRON

Manufacturer: Menley & James Laboratories, Inc.

Product Form: Tablet, Ch.

Ingredient	Concentration (Per Tablet)
Vitamin A (as Acetate)	1250 I.U.
Vitamin A (as Beta-Carotene)	1250 I.U.
Vitamin D	400 I.U.
Vitamin E	15 I.U.
Vitamin C	60mg
Folic Acid	0.3mg
Vitamin B_1 (as Thiamine)	1.05mg
Vitamin B_2 (Riboflavin)	1.2mg
Niacin	13.5mg
Vitamin B_6	1.05mg
Vitamin B_{12}	4.5mcg
Iron (Elemental)	15mg

All Ingredients:

Sucrose, Ascorbic Acid, Ferrous Fumarate, Mono- and Di-glycerides, Talc, Gelatin, Vitamin E Acetate, Niacinamide, Flavors, Starch, Calcium Carbonate, Colors, Silica, Magnesium Stearate, Stearic Acid, Pyridoxine Hydrochloride, Riboflavin, Thiamine Mononitrate, Beta-Carotene, Lactose, Vitamin A Acetate, Folic Acid, Vitamin D, Vitamin B_{12}.

GARFIELD, COMPLETE WITH MINERALS

Manufacturer:
Menley & James Laboratories, Inc.

Product Form:
Tablet, Ch.

Ingredient	Concentration (Per Tablet)	Ingredient	Concentration (Per Tablet)
Vitamin A (as Acetate)	2500 I.U.	Niacin	20mg
Vitamin A (as Beta-Carotene)	2500 I.U.	Vitamin B_6	2mg
Vitamin D	400 I.U.	Vitamin B_{12}	6mcg
Vitamin E	30 I.U.	Iron (Elemental)	18mg
Vitamin C	60mg	Biotin	40mcg
Folic Acid	0.4mg	Pantothenic Acid	10mg
Vitamin B_1 (as Thiamine)	1.5mg	Calcium	100mg
Vitamin B_2 (Riboflavin)	1.7mg	Copper	2mg
		Phosphorous	100mg
		Iodine	150mcg
		Magnesium	20mg
		Zinc	15mg

All Ingredients:
Dicalcium Phosphate, Sodium Ascorbate, Sorbitol, Ferrous Fumarate, Mono- and Di-glycerides, Gelatin, Magnesium Oxide, Microcrystalline Cellulose, Vitamin E Acetate, Talc, Zinc Oxide, Niacinamide, Flavors, Starch, Calcium Pantothenate, Citric Acid, Colors, Silica, Magnesium Stearate, Stearic Acid, Tartaric Acid, Pyridoxine Hydrochloride, Starch, Hydrogenated Cotton Seed Oil, Aspartame, Riboflavin, Thiamine Mononitrate, Beta-Carotene, Vitamin A Acetate, Folic Acid, Vitamin D, Vitamin B_{12}, Pyridoxine Hydrochloride, Cupric Oxide, Potassium Iodide, Biotin.

GERITOL

Manufacturer:
SmithKline Beecham, Inc.

Product Form:
Liquid

Ingredient:	Concentration: (Per 15ml)
Iron (as Ferric Ammonium Citrate)	50mg
Vitamin B_1 (as Thiamine)	2.5mg
Vitamin B_2 (as Riboflavin)	2.5mg
Niacinamide	50mg
Panthenol	2mg
Vitamin B_6 (as Pyridoxine)	0.5mg
Vitamin B_{12} (as Cyanocobalamin)	0.75mcg
Methionine	25mg
Choline Bitartrate	50mg

Other Ingredients:
Alcohol, Benzoic Acid, Color, Citric Acid, Invert Sugar, Sucrose, Water, Flavors.

GERITOL, COMPLETE

Manufacturer:
SmithKline Beecham, Inc.

Product Form:
Tablet

Ingredient	Concentration (Per Tablet)	Ingredient	Concentration (Per Tablet)
Vitamin A	6000 I.U.	Vitamin K	25mcg
Vitamin D	400 I.U.	Selenium	15mcg
Vitamin C	60mg	Calcium	162mg
Folic Acid	400mcg	Phosphorous	125mg
Vitamin B_1	1.5mg	Magnesium	100mg
Vitamin B_2	1.7mg	Zinc	15mg
Niacin	20mg	Iodine	150mcg
Vitamin B_6	2.0mg	Iron	50mg
Vitamin B_{12}	6mcg	Copper	2mg
Vitamin E	30 I.U.	Manganese	2.5mg
Biotin	45mcg	Potassium	37.5mg
Pantothenic Acid	10mg	Chloride	34mg
Chromium	15mcg	Molybdenum	15mcg
Vanadium	10mcg	Nickel	5mcg
Silicon	80mcg	Tin	10mcg

Other Ingredients:
Carnauba Wax, Crospovidone, Colors, Flavors, Gelatin, Glycerides of Stearic and Palmitic Acids, Hydroxypropylcellulose, Hydroxypropyl Methylcellulose, Magnesium Stearate, Microcrystalline Cellulose, Polyethylene Glycol, Silicon Dioxide, Stearic Acid, White Wax, Titanium Dioxide.

GERITOL, EXTEND

Manufacturer:
SmithKline Beecham, Inc.

Product Forms:
Tablet
Caplet

Ingredient	Concentration (Per Tablet or Caplet)
Vitamin A	3333 I.U.
Vitamin D	200 I.U.
Vitamin C	60mg
Folic Acid	0.2mg
Vitamin B_1	1.2mg
Vitamin B_2	1.4mg
Niacin	15mg
Vitamin B_6	2.0mg
Vitamin B_{12}	2mcg
Vitamin E	15 I.U.
Vitamin K	80mcg
Selenium	70mcg
Calcium	130mg
Phosphorous	100mg
Magnesium	35mg
Zinc	15mg
Iodine	150mcg
Iron	10mg

Other Ingredients:
Carnauba Wax, Croscarmellose Sodium, Flavors, Gelatin, Glycerides of Stearic and Palmitic Acid, Hydroxypropyl Methylcellulose, Magnesium Stearate, Microcrystalline Cellulose, Polyethylene Glycol, Silicon Dioxide, Stearic Acid, Colors, White Wax, Titanium Dioxide.

GEVRABON

Manufacturer:	Product Form:
Lederle Laboratories	Liquid
Active Ingredient:	**Concentration:** (Per 30ml)
Vitamin B_1 (as Thiamine Hydrochloride)	5mg
Vitamin B_2 (as Riboflavin-5-Phosphate Sodium	2.5mg
Niacinamide	50mg
Vitamin B_6 (Pyridoxine Hydrochloride)	1mg
Vitamin B_{12} (as Cyanocobalamin)	1mcg
Pantothenic Acid (as D-Pantothenyl Alcohol)	10mg
Iodine (as Potassium Iodide)	100mcg
Iron (as Ferrous Gluconate)	15mg
Magnesium (as Magnesium Chloride)	2mg
Zinc (as Zinc Chloride)	2mg
Choline (as Tricholine Citrate)	100mg
Manganese (as Manganese Chloride)	2mg

Other Ingredients:
Alcohol (18%), Citric Acid, Glycerin, Sherry Wine, Sucrose.

GEVRAL T

Manufacturer:
Lederle Laboratories

Product Form:
Tablet

Ingredient:	Concentration: (Per Tablet)
Vitamin A (as Acetate)	5000 I.U.
Vitamin E (as *dl*-Tocopheryl Acetate)	45 I.U.
Vitamin C (as Ascorbic Acid)	90mg
Vitamin B_1 (as Thiamine Mononitrate)	2.25mg
Folic Acid	0.4mg
Vitamin B_2 (as Riboflavin)	2.6mg
Niacinamide	30mg
Vitamin B_6 (Pyridoxine Hydrochloride)	3mg
Vitamin B_{12} (as Cyanocobalamin)	9mcg
Vitamin D_2	400 I.U.
Calcium (as Dibasic Calcium Phosphate)	162mg
Phosphorous (as Dibasic Calcium Phosphate)	125mg
Iodine (as Potassium Iodide)	225mcg
Iron (as Ferrous Fumarate)	27mg
Magnesium (as Magnesium Oxide)	100mg
Zinc (as Zinc Oxide)	22.5mg
Copper (as Cupric Oxide)	1.5mg

Other Ingredients:
BHA, BHT, Colors, Gelatin, Hydrolyzed Protein, Hydroxypropyl Methylcellulose, Lactose, Magnesium Stearate, Methylparaben, Microcrystalline Cellulose, Modified Food Starch, Mono- and Di-Glycerides, Polacrilin, Polysorbate 60, Potassium Sorbate, Propylparaben, PVPP, Silica Gel, Sodium Benzoate, Sodium Lauryl Sulfate, Sorbic Acid, Stearic Acid, Sucrose, Titanium Dioxide and other ingredients.

HALLS, DROPS

Manufacturer:
Warner-Lambert Company

Product Form:
Lozenge

Active Ingredient & Concentration
60mg Vitamin C/Lozenge

All Ingredients:
Sugar, Glucose Syrup, Sodium Ascorbate, Citric Acid, Ascorbic Acid, Flavors, Colors.

HYLAND'S, FOR CHILDREN

Manufacturer:
Standard Homeopathic Company

Product Form:
Tablet

Active Ingredient & Concentration:
25mg Vitamin C(as Sodium Ascorbate)/Tablet

Other Ingredients:
Lactose, Flavor.

MYADEC

Manufacturer: Parke-Davis

Product Form: Tablet

Active Ingredient	Concentration (Per Tablet)	Active Ingredient	Concentration (Per Tablet)
Vitamin A	5000 I.U.	Vitamin K	25mcg
Vitamin D	400 I.U.	Selenium	25mcg
Vitamin C	60mg	Calcium	162mg
Folic Acid	0.4mg	Phosphorous	125mg
Vitamin B_1	1.7mg	Magnesium	100mg
Vitamin B_2	2mg	Boron	150mcg
Niacin	20mg	Iodine	150mcg
Vitamin B_6	3mg	Iron	18mg
Vitamin B_{12}	6mcg	Copper	2mg
Vitamin E	30 I.U.	Manganese	2.5mg
Biotin	30mcg	Potassium	40mg
Pantothenic Acid	10mg	Chloride	36.3mg
Chromium	25mcg	Molybdenum	25mcg
Zinc	15mg	Nickel	5mcg
Silicon	10mcg	Tin	10mcg
Vanadium	10mcg		

All Ingredients:

Dicalcium Phosphate, Magnesium Oxide, Potassium Chloride, Ascorbic Acid, *dl*-Alpha Tocopheryl Acetate, Ferrous Fumarate, Modified Starch Gluconate, Starch, Cellulose, Modified Cellulose Gum, Niacinamide, Silica, Zinc Oxide, Polyethylene Oxide, *d*-Calcium Pantothenate, Beta-Carotene, Vitamin A Acetate, Manganese Sulfate, Magnesium Stearate, Pyridoxine Hydrochloride, Phytonadione, Biotin, Cupric Oxide, Riboflavin, Thiamine Mononitrate, Cyanocobalamin, Vitamin D, Folic Acid, Potassium Iodide, Sodium Potassium Borate, Chromium Chloride, Sodium Selenate, Sodium Molybdate, Stannous Chloride Sodium Metasilicate, Sodium Metavanadate, Nickelous Sulfate, Pharmaceutical Glaze, Methylcellulose, Polysorbate, Titanium Dioxide, Colors, Povidone, Polyethylene Glycol, Hydroxypropylcellulose.

ONE-A-DAY, MAXIMUM FORMULA

Manufacturer:
Miles, Inc.

Product Form:
Tablet

Active Ingredient	Concentration (Per Tablet)	Active Ingredient	Concentration (Per Tablet)
Vitamin A	5000 I.U.	Chromium	10mcg
Vitamin D	400 I.U.	Selenium	10mcg
Vitamin C	60mg	Calcium	130mg
Folic Acid	0.4mg	Phosphorous	100mg
Vitamin B_1	1.5mg	Magnesium	100mg
Vitamin B_2	1.7mg	Zinc	15mg
Niacin	20mg	Iodine	150mcg
Vitamin B_6	2mg	Iron (Elemental)	18mg
Vitamin B_{12}	6mcg	Copper	2mg
Vitamin E	30 I.U.	Manganese	2.5mg
Biotin	30mcg	Potassium	37.5mg
Pantothenic Acid	10mg	Chloride	34mg
Molybdenum	10mcg		

All Ingredients:
Dicalcium Phosphate, Magnesium Hydroxide, Cellulose, Potassium Chloride, Ascorbic Acid, Gelatin, Ferrous Fumarate, Starch, Zinc Sulfate, Vitamin E Acetate, Citric Acid, Niacinamide, Hydroxypropyl Methylcellulose, Magnesium Stearate, Calcium Pantothenate, Selenium Yeast, Color, PVP, Hydroxypropylcellulose, Manganese Sulfate, Silica, Copper Sulfate, Chromium Yeast, Molybdenum Yeast, Pyridoxine Hydrochloride, Riboflavin, Thiamine Mononitrate, Beta-Carotene, Vitamin A Acetate, Folic Acid, Potassium Iodide, Sodium Hexametaphosphate, Biotin, Vitamin D, Vitamin B_{12}, Lecithin.

ONE-A-DAY, WOMEN'S FORMULA

Manufacturer:
Miles, Inc.

Product Form:
Tablet

Active Ingredient:	Concentration (Per Tablet)
Vitamin A (as Acetate and Beta Carotene)	5000 I.U.
Vitamin C	60mg
Vitamin B_1 (Thiamine)	1.5mg
Vitamin B_2 (Riboflavin)	1.7mg
Niacin	20mg
Vitamin D	400 I.U.
Zinc	15mg
Vitamin E	30 I.U.
Vitamin B_6	2mg
Folic Acid	0.4mg
Vitamin B_{12}	6mcg
Pantothenic Acid	10mg
Iron (Elemental)	27mg
Calcium (Elemental)	450mg

Ingredients:
Calcium Carbonate, Acacia, Ferrous Fumarate, Ascorbic Acid, Gelatin, Vitamin E Acetate, Cellulose, Hydroxypropyl Methylcellulose, Niacinamide, Zinc Oxide, Modified Cellulose Gum, Calcium Pantothenate, Magnesium Stearate, Color, Hydroxypropylcellulose, Starch, Pyridoxine Hydrochloride, Riboflavin, Thiamine, Mononitrate, Beta-Carotene, Vitamin A Acetate, Folic Acid, Lecithin, Sodium Hexametaphosphate, Vitamin D, Vitamin B_{12}.

PROTEGRA

Manufacturer:
Lederle Laboratories

Product Form:
Liquigel

Ingredient:	Concentration: (per Liquigel)
Vitamin E	200 I.U.
Vitamin C	250mg
Beta-Carotene	3mg
Zinc	7.5mg
Copper	1mg
Selenium	15mcg
Manganese	1.5mg

Other Ingredients:
Gelatin, Cotton Seed Oil, Glycerin, Dibasic Calcium Phosphate, Lecithin, Partially Hydrogenated Cotton Seed and Soybean Oils, Beeswax, Titanium Dioxide FD&C Red No. 40, FD&C Yellow No. 6.

SIGTAB

Manufacturer:
Roberts Pharmaceutical Company

Product Form:
Tablet

Active Ingredient	Concentration (Per Tablet)	Active Ingredient	Concentration (Per Tablet)
Vitamin A	5000 I.U.	Vitamin E	15 I.U.
Vitamin C	333mg	Vitamin B_6	6mg
Vitamin B_1	10.3mg	Folic Acid	0.4mg
Vitamin B_2	10mg	Vitamin B_{12}	18mcg
Niacin	100mg	Pantothenic Acid	20mg
Vitamin D	400 I.U.		

All Ingredients:
Sucrose, Ascorbic Acid, Calcium Sulfate, Niacinamide, Vitamin E Acetate, Calcium Pantothenate, Vitamin A Acetate, Thiamine Mononitrate, Riboflavin, Gelatin, Pyridoxine Hydrochloride, Povidone, Lacca, Magnesium Stearate, Silica, Color, Sodium Benzoate, Folic Acid, Polyethylene Glycol, Cholecalciferol, Carnauba Wax, Medical Antifoam, Sesame Seed Oil, Titanium Dioxide, Cyanocobalamin.

SIGTAB-M

Manufacturer:	Product Form:
Roberts Pharmaceutical Corp.	Tablet

Active Ingredients:	Concentration (Per Tablet)
Vitamin A (as Vitamin A & Beta Carotene)	6000 I.U.
Vitamin D_3	400 I.U.
Vitamin E (*dl*-Tocopheryl Acetate)	45 I.U.
Vitamin C (Ascorbic Acid)	100mg
Niacinamide	25mg
Thiamine Mononitrate	5mg
Riboflavin	5mg
Pyridoxine Hydrochloride	3mg
Folic Acid	400mcg
Pantothenic Acid (D-Calcium Pantothenate)	15mcg
Vitamin B_{12}	18mcg
Vitamin K_1	25mcg
Biotin	45mcg
Calcium	200mg
Phosphorous (Dicalcium Phosphate)	150mg
Iron (Ferrous Fumarate)	18mg
Magnesium (Magnesium Oxide)	100mg
Copper (Copper Oxide)	2mg
Zinc (Zinc Oxide)	15mg
Iodine (Potassium Iodide)	150mcg
Manganese (Manganese Sulfate)	5mg
Potassium (Potassium Chloride)	40mg
Chloride	36.3mg
Selenium (Sodium Selenate)	25mcg
Molybdenum (Sodium Molybdate)	25mcg
Chromium (Chromium Chloride)	25mcg
Nickel	5mcg
Tin	10mcg
Vanadium	10mcg
Silicon	2mcg
Boron	150mcg

Other Ingredients:

Microcrystalline Cellulose, Stearic Acid, Croscarmellose Sodium, Magnesium Stearate, Hydroxypropyl Methylcellulose, Propylene Glycol, FD&C Yellow No. 6.

STRESSTABS

Manufacturer:
Lederle Laboratories

Product Form:
Tablet

Ingredient:	Concentration: (Per Tablet)
Vitamin E (as *dl*-Alpha Tocopheryl Acetate)	30 I.U.
Vitamin C (as Ascorbic Acid)	500mg
Folic Acid	400mcg
Vitamin B_1 (as Thiamine Mononitrate)	10mg
Vitamin B_2 (as Riboflavin)	10mg
Niacinamide	100mg
Vitamin B_6 (as Pyridoxine Hydrochloride)	5mg
Vitamin B_{12} (as Cyanocobalamin)	2mcg
Biotin	45mcg
Pantothenic Acid (as Calcium Pantothenate)	20mg

Other Ingredients:
Calcium Carbonate, FD&C Yellow No. 6, Magnesium Stearate, Microcrystalline Cellulose, Modified Food Starch, Silica Gel, Stearic Acid.

STRESSTABS, + IRON

Manufacturer:
Lederle Laboratories

Product Form:
Tablet

Ingredient:	Concentration: (Per Tablet)
Vitamin E (as *dl*-Alpha Tocopheryl Acetate)	30 I.U.
Vitamin C (as Ascorbic Acid)	500mg
Folic Acid	400mcg
Vitamin B_1 (as Thiamine Mononitrate)	10mg
Vitamin B_2 (as Riboflavin)	10mg
Niacinamide	100mg
Vitamin B_6 (as Pyridoxine Hydrochloride)	5mg
Vitamin B_{12} (as Cyanocobalamin)	12mcg
Biotin	45mcg
Pantothenic Acid (as Calcium Pantothenate)	20mg
Iron (as Ferrous Fumarate)	27mg

Other Ingredients:
Calcium Carbonate, Colors, Magnesium Stearate, Microcrystalline Cellulose, Modified Food Starch, Silica Gel, Stearic Acid.

STRESSTABS, + ZINC

Manufacturer: Lederle Laboratories

Product Form: Tablet

Ingredient:	Concentration: (Per Tablet)
Vitamin E (as *dl*-Alpha Tocopheryl Acetate)	30 I.U.
Vitamin C (as Ascorbic Acid)	500mg
Folic Acid	400mcg
Vitamin B_1 (as Thiamine Mononitrate)	10mg
Vitamin B_2 (as Riboflavin)	10mg
Niacinamide	100mg
Vitamin B_6	5mg
Vitamin B_{12} (as Cyanocobalamin)	12mcg
Biotin	45mcg
Pantothenic Acid (as Calcium Pantothenate)	20mg
Copper (as Cupric Oxide)	3mg
Zinc (as Zinc Sulfate)	23.9mg

Other Ingredients:
Calcium Carbonate, FD&C Yellow No. 6, Magnesium Stearate, Microcrystalline Cellulose, Modified Food Starch, Silica Gel, Stearic Acid.

STUART, PRENATAL

Manufacturer: Wyeth-Ayerst Laboratories

Product Form: Tablet

Active Ingredient	Concentration (Per Tablet)	Active Ingredient	Concentration (Per Tablet)
Vitamin A	4000 I/U.	Niacin	18mg
Vitamin D	400 I.U.	Vitamin B_6	2.6mg
Vitamin E	11mg	Vitamin B_{12}	4mcg
Vitamin C	100mg	Calcium	200mg
Folic Acid	0.8mg	Iron	60mg
Vitamin B_1	1.5mg	Zinc	25mg
Vitamin B_2	1.7mg		

All Ingredients:
Active: Calcium Sulfate, Ferrous Fumarate, Ascorbic Acid, *dl*-Tocopheryl Acetate, Zinc Oxide, Niacinamide, Vitamin A Acetate, Pyridoxine Hydrochloride, Riboflavin, Thiamine Mononitrate, Folic, Cholecalciferol, cyanocobalamin. Inactive: Croscarmellose Sodium, Hydroxypropyl Methylcellulose, Red Iron Oxide, Microcrystalline Cellulose, Pregelatinized Starch, Titanium Dioxide.

SUNKIST, CHILDREN'S REGULAR

Manufacturer:
CIBA Consumer Pharmaceuticals

Product Form:
Tablet, Ch.

Active Ingredient	Concentration	Active Ingredient	Concentration
Vitamin A	2500 I.U.	Niacinamide	13.5mg
Vitamin D_3	400 I.U.	Vitamin B_6	1.05mg
Vitamin E	15 I.U.	Vitamin B_{12}	4.5mcg
Vitamin C	60mg	Vitamin K_1	5mcg
Folic Acid	0.3mg	Vitamin B_2	1.2mg
Vitamin B_1	1.05mg		

All Ingredients:
Sorbitol, Flavor, Sodium Ascorbate, Ascorbic Acid, Starch, Stearic Acid, Vitamin E, Niacinamide, Carrageenan, Magnesium Stearate, Citric Acid, Aspartame, Calcium Silicate, Silica, Colors, Riboflavin, Thiamine, Folic Acid, Beta-Carotene, Vitamin D_3, Vitamin K_1, Vitamin B_6, Vitamin B_{12}. Contains Phenylalanine.

SUNKIST, CHILDREN'S + EXTRA C

Manufacturer:
CIBA Consumer Pharmaceuticals

Product Form:
Tablet, Ch.

Active Ingredient	Concentration	Active Ingredient	Concentration
Vitamin A	2500 I.U.	Niacinamide	13.5mg
Vitamin D_3	400 I.U.	Vitamin B_6	1.05mg
Vitamin E	15 I.U.	Vitamin B_{12}	4.5mcg
Vitamin C	250mg	Vitamin K_1	5mcg
Folic Acid	0.3mg	Vitamin B_2	1.2mg
Vitamin B_1	1.05mg		

All Ingredients:
Sorbitol, Sodium Ascorbate, Ascorbic Acid, Flavors, Starch, Stearic Acid, Vitamin E, Niacinamide, Aspartame, Magnesium Stearate, Colors, Riboflavin, Vitamin B_6, Thiamine, Folic Acid, Beta-Carotene, Vitamin D_3, Vitamin K_1, Vitamin B_{12}, Calcium Silicate, Silica. Contains Phenylalanine.

SUNKIST, CHILDREN'S PLUS IRON

Manufacturer:
CIBA Consumer Pharmaceuticals

Product Form:
Tablet, Ch.

Active Ingredient	Concentration (Per Tablet)
Vitamin A	2500 I/U.
Vitamin D_3	400 I.U.
Vitamin E	15 I.U.
Vitamin C	60mg
Folic Acid	0.3mg
Vitamin B_1	1.05mg
Niacinamide	13.5mg
Vitamin B_6	1.05mg
Vitamin B_{12}	4.5mcg
Vitamin B_2	1.2mg
Iron	15mg
Vitamin K_1	5mcg

All Ingredients:
Sorbitol, Flavors, Ferrous Fumarate, Sodium Ascorbate, Starch, Ascorbic Acid, Stearic Acid, Vitamin E, Niacinamide, Carrageenan, Magnesium Stearate, Aspartame, Calcium Silicate, Silica, Colors, Vitamin A, Riboflavin, Vitamin B_6, Thiamine, Folic Acid, Beta-Carotene, Vitamin D_3, Vitamin K_1, Vitamin B_{12}. Contains Phenylalanine.

SUNKIST, CHILDREN'S COMPLETE

Manufacturer:
CIBA Consumer Pharmaceuticals

Product Form:
Tablet, Ch.

Active Ingredient	Concentration (Per Tablet)	Active Ingredient	Concentration (Per Tablet)
Vitamin A	5000 I.U.	Vitamin K_1	10mcg
Vitamin D_3	400 I.U.	Iron	18mg
Vitamin E	30 I.U.	Magnesium	20mg
Vitamin C	60mg	Iodine	150mcg
Folic Acid	0.4mg	Zinc	10mg
Vitamin B_1	1.5mg	Manganese	1mg
Vitamin B_2	1.7mg	Calcium	100mg
Niacinamide	20mg	Phosphorous	78mg
Vitamin B_6	2mg	Copper	2mg
Vitamin B_{12}	6mcg	Biotin	40mcg
Pantothenic Acid	10mg		

All Ingredients:
Sorbitol, Dicalcium Phosphate, Flavors, Ferrous Fumarate, Stearic Acid, Carrageenan, Starch, Sodium Ascorbate, Vitamin E, Magnesium Oxide, Ascorbic Acid, Niacinamide, Citric Acid, Pantothenic Acid, Zinc Oxide, Magnesium Stearate, Colors, Aspartame, Silica, Calcium Silicate, Vitamin A, Manganese Sulfate, Vitamin B_6, Cupric Oxide, Riboflavin, Thiamine, Folic Acid, Beta-Carotene, Potassium Iodide, Biotin, Vitamin D_3, Vitamin K_1, Vitamin B_{12}. Contains Phenylalanine.

SUNKIST VITAMIN C

Manufacturer:	Product Form:
CIBA Consumer Pharmaceuticals	Tablet, Ch.

Ingredient:	Concentration:
Vitamin C	60, 250 or 500mg/Tablet

All Ingredients:
Fructose, Sodium Ascorbate, Ascorbic Acid, Flavors, Citrus Bioflavonoids, Sorbitol, Sucrose, Silica Hydrogel, Magnesium Stearate, Modified Food Starch, Dextrin, Lactose.

THERAGRAN

Manufacturer:	Product Form:
Bristol-Myers Products	Liquid

Ingredient:	Concentration: (Per 5ml)
Vitamin A	5000 I.U.
Vitamin D	400 I.U.
Vitamin C	200mg
Thiamine	10mg
Riboflavin	10mg
Niacin	100mg
Vitamin B_6	4.1mg
Vitamin B_{12}	5mcg
Pantothenic Acid	21.4mg

All Ingredients:
Purified Water, Sucrose, Glycerin, Propylene Glycol, Sodium Ascorbate, Niacinamide, Polysorbate 80, Ascorbic Acid, Thiamine Hydrochloride, d-Panthenol, Carboxymethylcellulose Sodium, Riboflavin-5-Phosphate Sodium, Flavors, Sodium Benzoate, Methylparaben, Pyridoxine Hydrochloride, Vitamin A Palmitate, Cholecalciferol, Ferric Ammonium Citrate, Cyanocobalamin.

THERAGRAN, ADVANCED FORMULA

Manufacturer:
Bristol-Myers Products

Product Form:
Tablet

Active Ingredient	Concentration (Per Tablet)
Vitamin A	5000 I.U.
Vitamin D_3	400 I.U.
Vitamin E	30 I.U.
Vitamin C	90mg
Folic Acid	400mcg
Vitamin B_1	3mg
Pantothenic Acid	10mg
Niacin	20mg
Vitamin B_6	3mg
Vitamin B_{12}	9mcg
Vitamin B_2	3.4mg
Biotin	30mcg

All Ingredients:
Lactose, Ascorbic Acid, Microcrystalline cellulose, Gelatin, *dl*-Alpha Tocopheryl Acetate, Niacinamide, Starch, Calcium Pantothenate, Sodium Caseinate, Hydroxypropyl Methylcellulose, Povidone, Sucrose, Pyridoxine Hydrochloride, Riboflavin, Silicon Dioxide, Magnesium Stearate, Thiamine Mononitrate, Vitamin A Acetate, Polyethylene Glycol, Triacetin, Stearic Acid, Titanium Dioxide, Annetto, Beta-Carotene, FD&C Red No. 40, Folic Acid, Biotin, Ergocalciferol, Cyanocobalamin.

THERAGRAN-M

Manufacturer:
Bristol-Myers Products

Product Form:
Tablet

Active Ingredient	Concentration (Per Tablet)	Active Ingredient	Concentration (Per Tablet)
Vitamin A	5000 I.U.	Potassium	7.5mg
Vitamin D	400 I.U.	Iron	27mg
Vitamin E	30 I.U.	Magnesium	100mg
Vitamin C	90mg	Iodine	150mcg
Folic Acid	400mcg	Zinc	15mg
Vitamin B_1	3mg	Manganese	5mg
Vitamin B_2	3.4mg	Calcium	40mg
Niacin	20mg	Phosphorous	31mg
Vitamin B_6	3mg	Copper	2mg
Vitamin B_{12}	9mcg	Chromium	15mcg
Pantothenic Acid	10mg	Molybdenum	15mcg
Biotin	30mcg	Selenium	10mcg
Chloride	7.5mg		

All Ingredients:
Magnesium Oxide, Dibasic Calcium Phosphate, Lactose, Ascorbic Acid, Ferrous Fumarate, Gelatin, *dl*-Alpha Tocopheryl Acetate, Crospovidone, Niacinamide, Hydroxypropyl Methylcellulose, Zinc Oxide, Povidone, Manganese Sulfate, Potassium Chloride, Starch, Calcium Pantothenate, Sodium Caseinate, Pyridoxine Hydrochloride, Cupric Sulfate, Magnesium Stearate, Sucrose, Silicon Dioxide, Riboflavin, Thiamine Mononitrate, Stearic Acid, Polyethylene Glycol, Triacetin, Vitamin A Acetate, FD&C Red No. 40, Potassium Citrate, Beta-Carotene, Titanium Dioxide, Folic Acid, Potassium Iodide, FD&C Blue No. 2, Chromic Chloride, odium Molybdate, Biotin, Sodium Selenate, Ergocalciferol, Cyanocobalamin.

THERAGRAN, STRESS FORMULA

Manufacturer:
Bristol-Myers Products

Product Form:
Tablet

Active Ingredient	Concentration (Per Tablet)
Vitamin B_1	15mg
Vitamin B_2	15mg
Vitamin B_6	25mg
Vitamin B_{12}	12mcg
Vitamin C	600mg
Vitamin E	30 I.U.
Niacin	100mg
Pantothenic Acid	20mg
Iron	27mg
Folic Acid	400mcg
Biotin	45mcg

All Ingredients:
Ascorbic Acid, Niacinamide, Ferrous Fumarate, Starch, Lactose, Pyridoxine Hydrochloride, Crospovidone, *dl*-Alpha Tocopheryl Acetate, Gelatin, Calcium Pantothenate, Riboflavin, Povidone, Thiamine Mononitrate, Hydroxypropyl Methylcellulose, Sodium Caseinate, Magnesium Stearate, Silicon Dioxide, Stearic Acid, Polyethylene Glycol, Triacetin, Titanium Dioxide, FD&C Red No. 40, FD&C Yellow No. 6, Folic Acid, Biotin, Cyanocobalamin.

THE STUART FORMULA

Manufacturer:
Johnson & Johnson • MERCK
Consumer Pharmaceuticals Company

Product Form:
Tablet

Active Ingredient	Concentration (Per Tablet)	Active Ingredient	Concentration (Per Tablet)
Vitamin A	5000 I.U.	Calcium	160mg
Vitamin D	400 I.U.	Iron	18mg
Vitamin E	15 I.U.	Magnesium	100mg
Vitamin C	60mg	Iodine	150mcg
Folic Acid	0.4mg	Phosphorous	125mg
Vitamin B_1	1.2mg	Vitamin B_6	2mg
Vitamin B_2	1.7mg	Vitamin B_{12}	6mcg
Niacin	20mg		

All Ingredients:
Active: Dibasic Calcium Phosphate, Magnesium Oxide, Ascorbic Acid, Ferrous Fumarate, *dl*-Alpha Tocopheryl Acetate, Folic Acid, Niacinamide, Vitamin A Palmitate, Cyanocobalamin, Pyridoxine Hydrochloride, Riboflavin, Thiamine Mononitrate, Ergocalciferol, Potassium Iodide. Other: Calcium Sulfate, Carnauba Wax, Pharmaceutical Glaze, Povidone, Sodium Starch Glycolate, Starch, Sucrose, Titanium Dioxide, White Wax.

TROPH-IRON

Manufacturer:
Menley & James Laboratories, Inc.

Product Form:
Liquid

Active Ingredient:	Concentration: (Per 5ml)
Vitamin B_1 (as Thiamine Hydrochloride)	10mg
Vitamin B_{12} (as Cyanocobalamin)	25mcg
Iron (as Ferric Pyrophosphate)	20mg

Other Ingredients:
Citric Acid, FD&C Red No. 40, Flavor, Glucose, Glycerin, Methylparaben, Propylparaben, Sodium Citrate, Saccharin Sodium, Purified Water.

TROPHITE

Manufacturer:
Menley & James Laboratories, Inc.

Product Form:
Liquid

Ingredient:	Concentration: (per 5ml)
Vitamin B_1 (as Thiamine Hydrochloride)	10mg
Vitamin B_{12} (as Cyanocobalamin)	25mcg

Other Ingredients:
Colors, Dextrose, Flavor, Glycerin, Methylparaben, Propylparaben, Sodium Tartrate, Tartaric Acid, Purified Water.

UNICAP

Manufacturer:
The Upjohn Company

Product Form:
Capsule

Active Ingredient	Concentration (Per Capsule)	Active Ingredient	Concentration (Per Capsule)
Vitamin A	5000 I.U.	Thiamine	1.5mg
Vitamin D	400 I.U.	Riboflavin	1.7mg
Vitamin E	30 I.U.	Vitamin B_6	2mg
Vitamin C	60mg	Vitamin B_{12}	6mcg
Folic Acid	400mcg	Niacin	20mg

All Ingredients:
Gelatin, Ascorbic Acid, Soybean Oil, Glycerin, Vitamin E Acetate, Niacinamide, Yellow Wax, Lecithin, Pyridoxin Hydrochloride, Thiamine Mononitrate, Riboflavin, Vitamin A Palmitate, Titanium Dioxide, Corn Oil, Folic Acid, FD&C Yellow No. 5, Ethyl Vanillin, Vanillin Enhancer, FD&C Yellow No. 6, Cholecalciferol, Cyanocobalamin.

UNICAP Jr.

Manufacturer:
The Upjohn Company

Product Form:
Tablet, Ch.

Active Ingredient	Concentration (Per Tablet)	Active Ingredient	Concentration (Per Tablet)
Vitamin A	5000 I.U.	Thiamine	1.5mg
Vitamin D	400 I.U.	Riboflavin	1.7mg
Vitamin E	15 I.U.	Vitamin B_6	2mg
Vitamin C	60mg	Vitamin B_{12}	6mcg
Folic Acid	400mcg	Niacin	20mg

All Ingredients:
Sodium Ascorbate, Sucrose, Mannitol, Lactose, Corn Starch, Citric Acid, Vitamin E Acetate, Niacinamide, Povidone, Flavor, Dextrins, Pyridoxin Hydrochloride, Thiamine Mononitrate, Riboflavin, Vitamin A Acetate, Folic Acid, Cholecalciferol, Cyanocobalamin, Silica, Calcium Stearate, Color.

UNICAP, PLUS IRON

Manufacturer:
The Upjohn Company

Product Form:
Tablet

Active Ingredient	Concentration (Per Tablet)	Active Ingredient	Concentration (Per Tablet)
Vitamin A	5000 I.U.	Thiamine	1.5mg
Vitamin D	400 I.U.	Riboflavin	1.7mg
Vitamin E	30 I.U.	Vitamin B_6	2mg
Vitamin C	60mg	Vitamin B_{12}	6mcg
Folic Acid	400mcg	Niacin	20mg
Pantothenic Acid	10mg	Calcium	100mg
Iron	22.5mg		

All Ingredients:
Calcium Phosphate, Cellulose, Ascorbic Acid, Ferrous Fumarate, Vitamin E Acetate, Color, Niacinamide, Vitamin A Acetate, Calcium Pantothenate, Magnesium Stearate, Silica Gel, Pyridoxine Hydrochloride, Riboflavin, Thiamine Mononitrate, Folic Acid, Flavor, Cholecalciferol, Carnauba Wax, Cyanocobalamin.

UNICAP M

Manufacturer:
The Upjohn Company

Product Form:
Tablet

Active Ingredient	Concentration (Per Tablet)	Active Ingredient	Concentration (Per Tablet)
Vitamin A	5000 I.U.	Thiamine	1.5mg
Vitamin D	400 I.U.	Riboflavin	1.7mg
Vitamin E	30 I.U.	Vitamin B_6	2mg
Vitamin C	60mg	Vitamin B_{12}	6mcg
Folic Acid	400mcg	Niacin	20mg
Pantothenic Acid	10mg	Calcium	60mg
Iron	18mg	Phosphorous	45mg
Iodine	150mcg	Manganese	1mg
Copper	2mg	Potassium	5mg
Zinc	15mg		

All Ingredients:
Calcium Phosphate, Ascorbic Acid, Vitamin E Acetate, Ferrous Fumarate, Cellulose, Niacinamide, Color, Zinc Oxide, Calcium Pantothenate, Vitamin A Acetate, Potassium Sulfate, Magnesium Stearate, Cupric Sulfate, Silica Gel, Manganese Sulfate, Pyridoxine Hydrochloride, Riboflavin, Thiamine Mononitrate, Folic Acid, Flavor, Cholecalciferol, Potassium Iodide, Carnauba Wax, Cyanocobalamin.

UNICAP Sr.

Manufacturer:
The Upjohn Company

Product Form:
Tablet

Active Ingredient	Concentration (Per Tablet)
Vitamin A	5000 I.U.
Vitamin D	200 I.U.
Vitamin E	15 I.U.
Vitamin C	60mg
Folic Acid	400mcg
Pantothenic Acid	10mg
Iron	10mg
Iodine	150mcg
Copper	2mg
Zinc	15mg
Thiamine	1.2mg
Riboflavin	1.4mg
Vitamin B_6	2.2mg
Vitamin B_{12}	3mcg
Niacin	16mg
Calcium	100mg
Phosphorous	77mg
Manganese	1mg
Potassium	5mg
Magnesium	30mg

All Ingredients:
Calcium Phosphate, Ascorbic Acid, Vitamin E Acetate, Ferrous Fumarate, Cellulose, Magnesium Oxide, Niacinamide, Color, Zinc Oxide, Calcium Pantothenate, Vitamin A Acetate, Potassium Sulfate, Magnesium Stearate, Cupric Sulfate, Silica Gel, Manganese Sulfate, Beta-Carotene, Pyridoxine Hydrochloride, Riboflavin, Thiamine Mononitrate, Folic Acid, Flavor, Cholecalciferol, Potassium Iodide, Carnauba Wax, Cyanocobalamin.

UNICAP T

Manufacturer:
The Upjohn Company

Product Form:
Tablet

Active Ingredient	Concentration (Per Tablet)
Vitamin A	5000 I.U.
Vitamin D	400 I.U.
Vitamin E	30 I.U.
Vitamin C	500mg
Folic Acid	400mcg
Pantothenic Acid	25mg
Iron	18mg
Iodine	150mcg
Copper	2mg
Zinc	15mg
Thiamine	10mg
Riboflavin	10mg
Vitamin B_6	6mg
Vitamin B_{12}	18mcg
Niacin	100mg
Manganese	1mg
Potassium	5mg
Selenium	10mcg

All Ingredients:
Ascorbic Acid, Niacinamide, Cellulose, Hydroxypropyl Methylcellulose, Vitamin E Acetate, Ferrous Fumarate, Colors, Calcium Pantothenate, Calcium Phosphate, Zinc Oxide, Vitamin A Acetate, Magnesium Stearate, Potassium Sulfate, Thiamine Mononitrate, Riboflavin, Selenium Yeast, Pyridoxine Hydrochloride, Cupric Sulfate, Silica Gel, Manganese Sulfate, Flavor, Folic Acid, Cholecalciferol, Potassium Iodide, Carnauba Wax, Cyanocobalamin.

Z-BEC

Manufacturer:
A.H. Robins Company, Inc.

Product Form:
Tablet

Active Ingredient	Concentration (Per Tablet)	Active Ingredient	Concentration (Per Tablet)
Vitamin E	45 I.U.	Vitamin B_6	10mg
Vitamin C	600mg	Vitamin B_{12}	6.0mcg
Vitamin B_1	15mg	Pantothenic Acid	25mg
Vitamin B_2	10.2mg	Zinc	22.5mg
Niacin	100mg		

Ingredients:

Niacinamide Ascorbate, Ascorbic Acid, Microcrystalline Cellulose, Vitamin E Acetate, Zinc Sulfate, Calcium Pantothenate, Hydroxypropyl Methylcellulose, Thiamine Mononitrate, Stearic Acid, Pyridoxine Hydrochloride, Riboflavin, Propylene Glycol, Titanium Dioxide, FD&C Blue No. 1 Aluminum Lake, Methylparaben, Propylparaben, Xanthan Gum, Sodium Citrate, Potassium Sorbate, Silicon Dioxide, Polysorbate 20, Magnesium Stearate, PVP, Vanillin, Cyanocobalamin.

ZYMACAP

Manufacturer:
Roberts Pharmaceutical Company

Product Form:
Capsule

Active Ingredient	Concentration (Per Capsule)	Active Ingredient	Concentration (Per Capsule)
Vitamin A	5000 I.U.	Vitamin D	400 I.U.
Vitamin E	15 I.U.	Vitamin B_6	3mg
Vitamin C	90mg	Vitamin B_{12}	9mcg
Vitamin B_1	2.25mg	Pantothenic Acid	15mg
Vitamin B_2	2.6mg	Folic Acid	400mcg
Niacin	30mg		

All Ingredients:

Soybean Oil, Ascorbic Acid, Gelatin, Glycerin, Niacinamide, Calcium Pantothenate, Vitamin E Acetate, Lecithin, Pyridoxine Hydrochloride, Yellow Wax, Thiamine Mononitrate, Riboflavin, Vitamin A Palmitate, Corn Oil, FD&C Red No. 40, Folic Acid, Titanium Dioxide, Ethyl Vanillin, Vanilla Enhancer, Cholecalciferol, Cyanocobalamin.

PART II

SUPPLIER'S FORMULATIONS

ASCORBIC ACID CHEWABLE TABLET, 100-1000mg

Supplier: BASF Corp.

Ingredient:	Wt.%
Ascorbic Acid, Powder	42.20
Sucrose, Ground	13.20
Sucrose, Crystalline	8.00
Microcrystalline Cellulose	28.30
Copolyvidone (**Kollidon VA64**)	2.40
PEG-6M	2.00
Flavors	1.20
Cyclamate Sodium	2.40
Saccharin Sodium	0.10
Silicon Dioxide	0.20
	100.00

Mixing Procedure:

Mix all the ingredients, pass them through a 0.8mm sieve and press to tablets weighing 250-2,500mg (corresponding to 100-1,000mg of the active ingredient).

BETA-CAROTENE TABLETS, 16mg

Supplier: BASF Corp.

Ingredient:	Wt.%
Beta-Carotene, Powder	39.22
Lactose(and)Povidone(and)Crospovidone (**Ludipress**)	58.82
Crospovidone	1.47
Magnesium Stearate	0.49
	100.00

Mixing Procedure:

Mix all the ingredients, pass them through a sieve and press into tablets with medium compressive force. Tablet weight = 400mg, Diameter = 12mm.

CALCIUM EFFERVESCENT TABLETS

Supplier: BASF Corp.

	Ingredient:	Wt.
	Calcium Lactate	650.00gm
	Calcium Gluconate	625.00gm
A	Calcium Carbonate	190.00gm
	Sodium Bicarbonate	410.00gm
	Tartaric Acid	480.00gm
	Povidone (**Kollidon 30**)	48.00gm
B	Povidone (**Kollidon 30**)	12.00gm
	Isopropanol or Ethanol, 99%	q.s.
	Crospovidone (**Kollidon CL**)	100.00gm
C	PEG-6M	48.00gm
	Flavor	q.s.

Mixing Procedure:

Granulate the Part A ingredients using the Part B solution. Dry, sieve and mix with the Part C ingredients. Press, using low compressive force, into tablets weighing 2,500mg with a diameter of 20mm.

MULTIVITAMIN EFFERVESCENT GRANULES

Supplier: BASF Corp.

	Ingredient:	Weight
	Thiamine Mononitrate	2.60gm
	Riboflavin	3.00gm
	Niacinamide	11.00gm
	Pyridoxine Hydrochloride	2.50gm
	Calcium-D-Pantothenate	15.00gm
	Ascorbic Acid	200.00gm
	Citric Acid	500.00gm
A	Sucrose	1,300.00gm
	Fructose	800.00gm
	Crospovidone (**Kollidon CL-M**)	200.00gm
	Cyclamate Sodium	1.00gm
	Saccharin Sodium	1.00gm
	Flavor	250.00gm
B	Povidone (**Kollidon 30**)	150.00gm
	Isopropanol or Ethanol	350.00gm
	Vitamin A Acetate, Powder 325,000CWD	15.00gm
	Vitamin D_3, Powder 100,000CWD	8.00gm
C	Vitamin E Acetate, Powder SD50	21.00gm
	Cyanocobalamin, 0.1% in Gelatin	6.60gm
	Sodium Bicarbonate	400.00gm

Mixing Procedure:

Granulate the Part A ingredients using the Part B solution in a fluidized bed granulator. Dry, sieve and mix with the Part C ingredients. Dry thoroughly before packaging under Nitrogen. 3-4gm = 1 RDA.

MULTIVITAMIN EFFERVESCENT TABLETS

Supplier: BASF Corp.

	Ingredient:	Weight
	Thiamine Mononitrate	13.00gm
	Riboflavin	4.00gm
	Niacinamide	66.00gm
	Pyridoxine Hydrochloride	11.00gm
A	Calcium-D-Pantothenate	17.00gm
	Tartaric Acid	360.00gm
	Sodium Bicarbonate	550.00gm
	Sucrose (Crystalline + Powder)	600.00gm
	Povidone (**Kollidon 30**)	35.00gm
B	Povidone (**Kollidon 30**)	5.00gm
	Isopropanol or Ethanol, 96%	~80.00gm
	Riboflavin	6.00gm
	Ascorbic Acid, Powder	550.00gm
	Vitamin A Palmitate,	
C	Powder 250,000 I.U./gm CWD	12.00gm
	Vitamin E Acetate, Powder SDG50	60.00gm
	Cyanocobalamin, 0.1% Gelatin coated	20.00gm
	PEG-6M	80.00gm
	Crospovidone (**Kollidon CL**)	100.00gm

Mixing Procedure:

Granulate the Part A ingredients using the Part B solution. Dry, sieve and mix with the Part C ingredients. Press into tablets weighing 2,500mg with a diameter of 20mm.

MULTIVITAMIN INSTANT GRANULES

Supplier: BASF Corp.

	Ingredient:	
	Vitamin A/D, Powder CWD	19.00gm
	Thiamine Mononitrate	2.60gm
	Riboflavin	3.30gm
	Niacinamide	11.00gm
	Pyridoxine Hydrochloride	2.20gm
	Cyanocobalamin, 0.1% Gelatin coated	6.60gm
	Ascorbic Acid, Powder	115.00gm
	Tocopheryl Acetate, Powder SD50	21.00gm
	Sucrose, Ground	2,000.00gm
	Crospovidone (**Kollidon CL-M**)	500.00gm
	Flavor	100.00gm
	Calcium-D-Pantothenate	150.00gm
B	Povidone (**Kollidon 30**)	200.00gm
	Ethanol, 96%	700.00ml

Mixing Procedure:

Granulate the Part A ingredients with the Part B solution using a fluidized bed granulator. If possible, the granules should be packed under Nitrogen.

MULITVITAMIN SYRUP

Supplier: BASF Corp.

	Ingredient:	Weight
	Vitamin A Palmitate, $1.7x10^6$ I.U./gm	10.00mg
	Vitamin D, $40x10^6$ I.U./gm	0.05mg
A	Vitamin E Acetate	100.00mg
	BHT	2.00mg
	PEG-40 Hydrogenated Castor Oil (**Cremophor RH 40**)	4.50gm
B	Water	10.00gm
	Sucrose	45.00gm
C	Methylparaben	200.00mg
	Citric Acid	80.00mg
D	Glycerin	10.00gm
	Water	25.00gm
	Thiamine Hydrochloride	15.00mg
	Riboflavin-5-Phosphate Sodium	15.00mg
	Niacinamide	55.00mg
E	Pyridoxine Hydrochloride	15.00mg
	Ascorbic Acid	0.30gm
	Sorbic Acid	0.10gm
	Propylene Glycol	5.14gm

Mixing Procedure:

Mix and heat the Part A ingredients to ~60°C. In a separate vessel, heat Part B to ~60°C. Mix Parts A and B slowly until a clear solution is obtained. Dissolve the Part C ingredients in a hot mixture of the Part D ingredients. Mix until a clear solution is obtained. Cool the solutions and mix Parts A+B with Parts C+D. Add the Part E ingredients and adjust the pH to 4.0-4.2. Bubble Nitrogen through the solution for 10 minutes and package under nitrogen.

DAILY VITAMIN TABLET

Supplier: FMC Corp.

	Ingredient:	Wt.%
	Folic Acid	0.15
A	Vitamin B_1	0.51
	Vitamin B_2	0.55
	Vitamin B_6	0.79
	Vitamin A + Vitamin D, 500A/40D	3.24
	Vitamin E, 50%, Acetate, Dry	9.26
	Vitamin C, 90%	22.65
	Niacinamide	6.47
B	Vitamin B_{12}, 0.1% Spray-Dried	2.65
	Dicalcium Phosphate	27.62
	Croscarmellose Sodium (**Ac-Di-Sol®**)	1.29
	Microcrystalline Cellulose (**Avicel® PH-101**)	21.18
	Ethyl Vanillin	0.05
C	Stearic Acid, XXX*	3.09
	Magnesium Stearate	0.50
		100.00

Mixing Procedure:

Prepare a premix of all the Part A ingredients. Blend the premix with all the Part B ingredients for 10 minutes. Screen in Stearic Acid and blend for 5 minutes. Screen in Magnesium Stearate and blend for an additional 5 minutes. Compress using 11/32" S.C. tooling.

---

* XXX is a common abbreviation for triple-pressed Stearic Acid

DECAVITAMIN TABLET

Supplier: FMC Corp.

	Ingredient:	Wt.%
	Folic Acid	0.07
	Niacinamide	10.28
A	Vitamin B_6, Pyridoxine Hydrochloride	1.03
	Vitamin B_2	1.03
	Vitamin B_1, Thiamine Hydrochloride	1.07
	Microcrystalline Cellulose (**Avicel PH-101**)	9.82
	Vitamin A + Vitamin D, 500A/50D	4.67
	Vitamin C, 90% Granular	39.72
	Calcium Pantothenate	6.07
B	Vitamin B_{12}, 0.1% Spray-Dried	3.50
	Vitamin E	14.72
	Croscarmellose Sodium (**Ac-Di-Sol**)	1.23
	Fumed Silica	0.49
	Dicalcium Phosphate	4.04
C	Stearic Acid, XXX	1.96
	Magnesium Stearate	0.30
		100.00

Mixing Procedure:

Prepare a premix of all the Part A ingredients. Blend the premix with all the Part B ingredients. Screen in the Stearic Acid and blend for 5 minutes. Screen in the Magnesium Stearate and blend an additional 5 minutes. Compress using 5/16" deep cup tooling.

GERIATRIC VITAMIN TABLET (SUGAR COATED)

Supplier: FMC Corp.

	Ingredient:	Wt.%
A	Vitamin B_6, Pyridoxine Hydrochloride	0.13
	Dicalcium Phosphate (**CalStar® DCP·$2H_2O$**)	33.36
	Iron, 95% Ferrous Sulfate/Ethylcellulose Granulation	29.46
	Vitamin B_1, Thiamine Mononitrate	1.03
	Vitamin B_2, Riboflavin	0.98
B	Niacinamide, Free Flow	5.89
	Vitamin C, Ascorbic Acid, 90%	17.86
	Calcium Pantothenate	0.72
	Croscarmellose Sodium (**Ac-Di-Sol SD-711**)	2.00
	Vitamin B_{12}, 0.1% Trituration	1.07
	Microcrystalline Cellulose (**Avicel PH-101**)	5.00
C	Stearic Acid XXX	2.00
	Magnesium Stearate	0.50
		100.00

Mixing Procedure:

Prepare a 5% trituration of the Part A ingredients. Mix the trituration with the Part B ingredients and blend for 10 minutes. Screen in the Stearic Acid and blend for 5 minutes. Screen in the Magnesium Stearate and mix for an additional 5 minutes. Compress using 0.3" with football shaped tooling.

MULTIVITAMIN TABLET WITH IRON (FILM COATED)

Supplier: FMC Corp.

	Ingredient:	Wt.%
	Folic Acid	0.14
A	Vitamin B_2	0.53
	Vitamin B_6	0.77
	Vitamin A + Vitamin D, 500A/40D	3.14
	Vitamin E, 50%, Acetate	9.00
	Vitamin B_1	0.50
	Vitamin C, 90%	22.00
	Niacinamide, Free Flow	6.29
B	Vitamin B_{12}, 0.1% Spray-Dried	0.73
	Iron	16.00
	Croscarmellose Sodium (**Ac-Di-Sol**)	1.53
	Microcrystalline Cellulose (**Avicel PH-101**)	22.12
	Ethyl Vanillin	0.05
	Dicalcium Phosphate (**Calstar DCP·2H_2O**)	13.70
C	Stearic Acid, XXX	3.00
	Magnesium Stearate	0.50
		100.00

Mixing Procedure:

Prepare a premix of all the Part A ingredients. Blend the premix with all the Part B ingredients for 10 minutes. Screen in the Stearic Acid and blend for 5 minutes. Screen in the Magnesium Stearate and blend an additional 5 minutes. Compress using 11/32" deep cup tooling.

MULTIVITAMIN TABLET WITH MINERALS

Supplier: FMC Corp.

	Ingredient:	Wt. %
	Thiamine Hydrochloride	0.13
	Riboflavin	0.14
	Pyridoxine Hydrochloride	0.21
	Cyanocobalamin, 1% Triturated, Spray-Dried	0.69
A	Folic Acid	0.04
	Iodine, as Potassium Iodide	0.02
	Copper, as dry Cupric Sulfate	0.43
	Microcrystalline Cellulose (**Avicel PH-101**)	17.57
	Vitamin A + Vitamin D, 500A/40D	0.85
	Ascorbic Acid, 90%	5.92
	Vitamin E Acetate, 50%, dry	2.42
	Niacinamide	1.69
	Pantothenic Acid, as Calcium Pantothenate	1.26
B	Iron, as Ferrous Fumarate	4.31
	Calcium, as Dicalcium Phosphate	42.77
	Magnesium Oxide	12.77
	Zinc, as Zinc Sulfate	3.23
	Croscarmellose Sodium (**Ac-Di-Sol SD-711**)	3.00
	Ethyl Vanillin	0.05
C	Stearic Acid, XXX	2.00
	Magnesium Stearate	0.50
		100.00

Mixing Procedure:

Prepare a premix of all the Part A ingredients. Blend the premix with all the Part B ingredients for 10 minutes. Screen in the Stearic Acid and blend for 5 minutes. Screen in the Magnesium Stearate and blend an additional 5 minutes. Compress using football shaped tooling.

STRESS FORMULA 600 WITH IRON

Supplier: FMC Corp.

	Ingredient:	Wt.%
	Iron, Ferrous Fumarate	7.00
	Vitamin B_1, Thiamine Mononitrate	1.50
	Vitamin B_2, Riboflavin	1.38
	Vitamin B_6, Pyridoxine Hydrochloride	2.29
	Vitamin B_{12}, Cyanocobalamin, 0.1%	1.50
	Vitamin C, Ascorbic Acid, 90%	61.08
A	Vitamin E, 50% Acetate	5.25
	Niacinamide	9.17
	Calcium Pantothenate, Granular	2.50
	Folic Acid	0.04
	Croscarmellose Sodium (**Ac-Di-Sol**)	2.50
	Fumed Silica	0.20
	Ethyl Vanillin	0.05
	Microcrystalline Cellulose (**Avicel PH-101**)	2.04
B	Stearic Acid, XXX	3.00
	Magnesium Stearate	0.50
		100.00

Mixing Procedure:

Blend all ingredients all the Part A ingredients for 10 minutes. Screen in Stearic Acid and blend for 5 minutes. Screen in Magnesium Stearate and blend for an additional 5 minutes. Compress using capsule shaped tooling.

STRESS FORMULA 600 TABLET

Supplier: FMC Corp.

	Ingredient:	Wt.%
	Vitamin B_1, Thiamine Mononitrate	1.57
	Vitamin B_2, Riboflavin	1.50
	Vitamin B_6, Pyridoxine Hydrochloride	0.52
	Vitamin B_{12}, Cyanocobalamin, 0.1%, Spray-Dried	1.64
	Vitamin C, Ascorbic Acid	63.64
A	Vitamin E, 50% Acetate	5.73
	Niacinamide	9.54
	Calcium Pantothenate, Granular	2.27
	Fumed Silica	0.20
	Ethyl Vanillin	0.05
	Microcrystalline Cellulose (**Avicel PH-101**)	10.00
	Dicalcium Phosphate, Unmilled	0.84
B	Stearic Acid, XXX	2.00
	Magnesium Stearate	0.50
		100.00

Mixing Procedure:

Blend all ingredients all the Part A ingredients for 10 minutes. Screen in Stearic Acid and blend for 5 minutes. Screen Magnesium Stearate and blend for an additional 5 minutes. Compress using capsule shaped tooling.

VITAMIN A TABLET, 10,000 UNITS

Supplier: FMC Corp.

	Ingredient:	Wt.%
	Vitamin A, as Acetate or Palmitate	10.00
A	Microcrystalline cellulose (**Avicel PH-101**)	10.91
	Croscarmellose Sodium (**Ac-Di-Sol**)	1.00
	Dicalcium Phosphate (**Calstar DCP ·2H$_2$O**)	75.59
B	Stearic Acid, XXX	2.00
	Magnesium Stearate	0.50
		100.00

Mixing Procedure:

Blend all the Part A ingredients for 10 minutes. Screen in the Stearic Acid and blend for 5 minutes. Screen in the Magnesium Stearate and blend for an additional 5 minutes. Compress using 5/16" S.C. tooling.

VITAMIN B$_1$ TABLET, 100mg

Supplier: FMC Corp.

	Ingredient:	Wt.%
	Thiamine Hydrochloride	22.00
A	Microcrystalline cellulose (**Avicel PH-102**)	10.00
	Croscarmellose Sodium (**Ac-Di-Sol**)	1.00
	Dicalcium Phosphate (**Calstar DCP•2H$_2$O**)	65.25
	Fumed Silica	0.25
B	Stearic Acid XXX	1.00
	Magnesium Stearate	0.50
		100.00

Mixing Procedure:

Blend all the Part A ingredients for 10 minutes. Screen in the Stearic Acid and blend for 5 minutes. Screen in the Magnesium Stearate and blend for an additional 5 minutes. Compress using 13/32" S.C. tooling.

VITAMIN B_2 TABLET, 50mg

Supplier: FMC Corp.

	Ingredient:	Wt.%
	Riboflavin	11.00
A	Microcrystalline cellulose (**Avicel PH-101**)	10.00
	Croscarmellose Sodium (**Ac-Di-Sol**)	1.00
	Dicalcium Phosphate (**Calstar DCP•$2H_2O$**)	74.40
B	Stearic Acid XXX	3.00
	Magnesium Stearate	0.60
		100.00

Mixing Procedure:

Blend all the Part A ingredients for 10 minutes. Screen in the Stearic Acid and blend for 5 minutes. Screen in the Magnesium Stearate and blend for an additional 5 minutes. Compress using 7/16" S.C. tooling.

VITAMIN B_6 TABLET, 50mg

Supplier: FMC Corp.

	Ingredient:	Wt.%
	Pyridoxine Hydrochloride	13.75
A	Microcrystalline cellulose (**Avicel PH-101**)	7.50
	Croscarmellose Sodium (**Ac-Di-Sol**)	1.00
	Dicalcium Phosphate (**Calstar DCP·$2H_2O$**)	75.55
	Fumed Silica	0.20
B	Stearic Acid XXX	1.50
	Magnesium Stearate	0.50
		100.00

Mixing Procedure:

Blend all the Part A ingredients for 10 minutes. Screen in the Stearic Acid and blend for 5 minutes. Screen in the Magnesium Stearate and blend for an additional 5 minutes. Compress using 3/8" S.C. tooling.

VITAMIN B_6 TABLET, 100mg

Supplier: FMC Corp.

	Ingredient:	Wt.%
	Pyridoxine Hydrochloride	20.00
A	Microcrystalline cellulose (**Avicel PH-101**)	10.00
	Croscarmellose Sodium (**Ac-Di-Sol**)	1.00
	Dicalcium Phosphate (**Calstar DCP•$2H_2O$**)	65.27
	Fumed Silica	0.27
B	Stearic Acid, XXX	2.91
	Magnesium Stearate	0.55
		100.00

Mixing Procedure:

Blend all the Part A ingredients for 10 minutes. Screen in the Stearic Acid and blend for 5 minutes. Screen in the Magnesium Stearate and blend for an additional 5 minutes. Compress using 13/32" S.C. tooling.

VITAMIN B_{12} TABLET, 100mg

Supplier: FMC Corp.

	Ingredient:	Wt.%
	Vitamin B_{12}, 1% Trituration in anhydrous Dicalcium Phosphate	2.78
A	Microcrystalline cellulose (**Avicel PH-101**)	10.00
	Croscarmellose Sodium (**Ac-Di-Sol**)	1.00
	Dicalcium Phosphate (**Calstar DCP•$2H_2O$**)	84.40
	Fumed Silica	0.22
B	Stearic Acid, XXX	1.11
	Magnesium Stearate	0.44
	FD&C Red #3 Lake	0.05
		100.00

Mixing Procedure:

Blend all the Part A ingredients for 10 minutes. Screen in the Stearic Acid and red color, through a 60 mesh sieve, and blend for 5 minutes. Screen in the Magnesium Stearate and blend for an additional 5 minutes. Compress using 13/32" S.C. tooling.

VITAMIN B_{12} TABLET, 500mg

Supplier: FMC Corp.

	Ingredient:	Wt.%
	Vitamin B_{12}, 1% Trituration in anhydrous Dicalcium Phosphate	13.33
A	Microcrystalline cellulose (**Avicel PH-101**)	14.89
	Croscarmellose Sodium (**Ac-Di-Sol**)	1.00
	Dicalcium Phosphate (**Calstar DCP•$2H_2O$**)	67.87
	Fumed Silica	0.22
B	Stearic Acid, XXX	2.22
	Magnesium Stearate	0.45
	FD&C Red #3 Lake	0.02
		100.00

Mixing Procedure:

Blend all the Part A ingredients for 10 minutes. Screen in the Stearic Acid and red color, through a 60 mesh sieve, and blend for 5 minutes. Screen in the Magnesium Stearate, through a 60 mesh sieve, and blend for an additional 5 minutes. Compress using 13/32" S.C. tooling.

MINERAL HARD GELATIN CAPSULE

Supplier: Gattefossé s.a.

	Ingredient:	Wt.%
A	Magnesium Chloride	42.86
B	**Gelucire 44/14**	57.14
		100.00

Mixing Procedure:

Melt Part B at 50°C. Allow it to cool to 40°C and mix in Part A. Pour this mixture into No. 2 hard gelatin capsules, 0.350gm/capsule

---

VITAMIN HARD GELATIN CAPSULE

Supplier: Gattefossé s.a.

	Ingredient:	Wt.%
A	Wheat Germ Oil	47.37
B	Vitamin E	0.79
C	**Gelucire 50/02**	51.84
		100.00

Mixing Procedure:

Melt Part C at 50°C. Allow it to cool to 40°C and mix in Parts A and B. Pour the mixture into No. 2 hard gelatin capsules, 0.380gm/capsule.

VITAMIN C TABLETS, 150mg

Supplier: Grain Processing Corp.

	Ingredient:	Wt.%
A	Maltodextrin (**Matrin M150**)	76.50
B	Vitamin C, Fine Granular	20.00
C	Silicon Dioxide	2.00
D	Magnesium Stearate	1.50
		100.00

Mixing Procedure:

Screen Part C through a U.S. standard 20 mesh screen. Dry blend Parts A, B and C together thoroughly. Screen Part D through a U.S. standard 20 mesh screen, add it to Parts A+B+C and mix. Tablets are prepared on a Stokes 16-station rotary press with 7/16" standard round tooling.

ASCORBIC ACID 500mg TABLET

Supplier: International Specialty Products

Ingredient:	Wt.%
Ascorbic Acid, 95%	87.00
Crospovidone (**Polyplasdone® XL or XL-10**)	2.00
Ditab	10.75
Magnesium Stearate	0.25
	100.00

Mixing Procedure:
Contact supplier for mixing procedure.

ONE-A-DAY TYPE VITAMIN FORMULA

Supplier: International Specialty Products

Ingredient:	Wt.%
Ditab	51.298
Ascorbic Acid, 95%	22.290
Vitamin E, 50% SD	9.980
Vitamin A, 5000 I.U.	3.330
Vitamin D, 400 I.U.	0.330
Crospovidone (**Polyplasdone XL or XL-10**)	4.000
Vitamin B_1	0.500
Vitamin B_2	0.570
Vitamin B_6	0.670
Vitamin B_{12}	0.002
Niacin	6.650
Folic Acid	0.130
Magnesium Stearate	0.250
	100.000

Mixing Procedure:
Contact supplier for mixing procedure.

VITAMIN B_{12} TABLETS

Supplier: International Specialty Products

	Ingredient:	Per Tablet
A	Calcium Sulfate	110.00mg
	Sucrose	20.00mg
B	Vitamin B_{12}, Cyanocobalamin	55.00mg
	Plasdone, 10% Solution (**PVP® K-30**)	q.s.
	Starch	10.00mg
C	Microcrystalline Cellulose	12.00mg
	Sterotex	0.75mg

Mixing Procedure:

Mix the Part A ingredients. Dissolve the Vitamin B_{12} in 50% of the Plasdone solution and add it slowly to Part A. Slowly add enough of the remaining Part B solution to granulate the mixture. Mix for 30 minutes. Pass the wet mass through a 14 mesh screen and dry the granules in an oven at 120-130°F. When dry, pass the mix through a 20 mesh screen. Mix the Part C ingredients and pass them through an 80 mesh screen. Add the Parts A+B to Part C and mix 20 minutes. Compress to weight with 1/4" standard concave punches.

ASCORBIC ACID 250mg TABLET

Supplier: R.T. Vanderbilt Company, Inc.

	Ingredient:	Wt.%
	Ascorbic Acid	41.67
	Magnesium Aluminum Silicate (**Veegum F**)	5.00
A	Microcrystalline Cellulose	20.00
	Starch	10.00
	Emdex	20.33
B	Stearic Acid	1.00
	Fumed Silica	2.00
		100.00

Mixing Procedure:

Blend the Part A ingredients until lump free (15 minutes). Add the Part B ingredients and blend for an additional 10 minutes. Compress into tablets with a mean weight of 586mg.

APPENDIX A

OTC PHARMACEUTICAL MANUFACTURERS

(Represented in this book)

-A-

Advanced Polymer Systems
3696 Haven Ave.
Redwood City, CA 94063

Allergan Pharmaceuticals,
Division of Allergan, Inc.
2525 DuPont Dr.
P.O. Box 19534
Irvine, CA 92713-9534

Almay, Inc.,
Division of Revlon, Inc.
625 Madison Ave.
New York, NY 10022

Apothecon,
A Bristol-Myers Squibb Co.
P.O. Box 4500
Princeton, NJ 08543-4500

B.F. Ascher & Company, Inc.
15501 West 109th St.
Lenexa, KS 66219
Mailing Address:
P.O. Box 717
Shawnee Mission, KS
66201-0717

Astra Pharmaceutical
Products, Inc.
50 Otis St.
Westboro, MA 01581-4500

Au Pharmaceuticals, Inc.
P.O. Box 131835
Tyler, TX 75713-1835

-B-

Baker Cummins
Dermatologicals, Inc.
1950 Swarthmore Ave.
Lakewood, NJ 10017-4071

Beach Pharmaceuticals,
Division of Beach
Products, Inc.
5220 S. Manhattan Ave.
Tampa, FL 33611

Beiersdorf, Inc
P.O. Box 5529
Norwalk, CT 06856-5529

Biopractic Group II, Inc.
99 Broad St.
P.O. Box 5300
Phillipsburg, NJ 08865

Block Drug Company, Inc.
257 Cornelison Ave
Jersey City, NJ 07302

Bock Pharmacal Company
P.O. Box 8519
St. Louis, MO 63126-0519

Bristol-Myers Products,
A Bristol-Myers Squibb Company
345 Park Ave.
New York, NY 10154

Burroughs Wellcome Company
3030 Cornwallis Rd.
Research Triangle Park, NC 27709

-C-

Carter-Wallace, Inc.
1345 Avenue of the Americas
New York, NY 10105

Chattem Consumer Products,
Division of Chattem, Inc.
1715 West 38th St.
Chattanooga, TN 37409

Chesebrough-Pond's, Inc.
33 Benedict Pl.
Greenwich, Ct 06830

CIBA Consumer Pharmaceuticals, Division of Ciba-Geigy Corp.
Mack Woodbridge II
Woodbridge, NJ 07095

Colombia Laboratories, Inc.
4000 Hollywood Blvd.
Hollywood, FL 33021

Combe, Inc.
1101 Westchester Ave.
White Plains, NY 10604

-D-

Del Pharmaceuticals, Inc.
163 E. Bethpage Rd.
Plainview, NY 11803

-F-

Fisons Consumer Health,
Division of Fisons Corp.
P.O. Box 1212
Rochester, NY 14603

Fleming & Company
1600 Fenpark Dr.
Fenton, MO 63026

-G-

The Gillette Company
Prudential Tower Building
Boston, MA, 02199

-H-

Helene Curtis Industries
325 N. Wells St,
Chicago, IL 60610

-J-

Johnson & Johnson Consumer
Products, Inc.
Grandview Rd.
Skillman, NJ 08558

Johnson & Johnson • Merck
Consumer Pharmaceuticals
Company
Camp Hill Rd,
Fort Washington, PA 19034

-K-

Konsyl Pharmaceuticals, Inc.
4200 S. Hulen
Fort Worth, TX 76109

-L-

Lederle Laboratories, Division
of American Cyanamid Company
One Cyanamid Plaza
Wayne, NJ 07470

-M-

McNeil Consumer Products
Company, Division Of
PCC, Inc.
Fort Washington, PA, 19034

Menley & James
Laboratories, Inc.
Commonwealth Corporate
Center
100 Tournament Dr,
Suite 310
Horsham, PA 19044-3697

The Mennen Company,
Division Of Colgate-McNeil-
Palmolive Co.
300 Park Ave.
New York, NY 10022-7499

Miles, Inc.
P.O. Box 340
Elkhart, IN 46515

Muro Pharmaceutical, Inc.
890 East Street
Tewksbury, MA 01876-1496

-N-

Neutrogena Corp.
5760 W. 96th St.
Los Angeles, CA 99045

-O-

Ohm Laboratories, Inc
P.O. Box 279
Franklin Park, NJ 08823

Ortho Pharmaceutical Corp.
Advanced Care Products
Raritan, NJ 08869

-P-

Parke-Davis, Division of
Warner-Lambert Company
201 Tabor Rd.
Morris Plains, NJ 07950

Pfizer Consumer Health
Care, Division of
Pfizer, Inc.
100 Jefferson Rd.
Parsippany, NJ 07054

The Procter & Gamble Co.
P.O. Box 5516
Cincinnati, OH 45201

-R-

Reed & Carnrick,
Division of Block Drug Co.
257 Cornelison Ave.
Jersey City, NJ 07302-9988

The Reese Chemical Co.
10617 Frank Ave.
Cleveland, OH 44106

Revlon Consumer Products,
Inc.
625 Madison Ave.
New York, NY 10022

Rhône-Poulenc Rorer
Pharmaceuticals, Inc.
500 Arcola Rd.
P.O. Box 1200
Collegeville, PA 19426

Roberts Pharmaceutical
Corp.
6 Industrial Way West
Eatontown, NJ 07724

A.H. Robbins Company, Inc,
Subsidiary of American Home
Products Corp.
1405 Cummings Dr.
Richmond, VA 23220

Ross Laboratories
6600 Bush Blvd.
Columbus, OH 43215

Rydelle Laboratories,
Division of S.C Johnson &
Son, Inc.
1525 Howe St.
Racine, WI 53403

-S-

Sandoz Pharmaceuticals/
Consumer Division
59 Route 10
East Hanover, NJ 07936

Sanofi Winthrop
Pharmaceuticals
90 Park Ave,
New York, NY 10016

Schering-Plough
HealthCare Products
Liberty Corner, NJ 07938

SmithKline Beecham Consumer
Brands, Division of
SmithKline Beecham, Inc.
P.O. Box 1467
Pittsburgh, PA 15230

Standard Homeopathic Co.
210 W. 131st St.
P.O. Box 61067
Los Angeles, CA 90061

Sterling Health, Division
of Sterling Winthrop, Inc.
90 Park Ave.
New York, NY 10016

-T-

Tanning Research, Inc.
1190 U.S. Rte. 1
Ormond Beach, FL 32074

Thompson Medical Co., Inc.
222 Lakeview Ave.
West Palm Beach, FL 33401

Triton Consumer Products,
Inc.
561 W. Golf Rd.
Arlington, IL 60005

-U-

The Upjohn Company
Kalamazoo, MI 49001

-W-

Wallace Laboratories
P.O. Box 1001
Half Acre Rd.
Cranbury, NJ 08512

Warner-Lambert Company
201 Tabor Rd.
Morris Plains, NJ 07950

Whitehall Laboratories,
Division of American Home
Products Corp,
685 Third Ave.
New York, NY 10017

Wyeth-Ayerst Laboratories,
Division of American Home
Products Corp.
P.O. Box 8299
Philadelphia, PA 19101

APPENDIX B

RAW MATERIAL SUPPLIERS

(Represented in this book)

-A-

Amerchol Corp.
P.O. Box 4051
136 Talmadge Rd.
Edison, NJ 08818
(908) 248-6000
FAX:(908) 287-4186

American Distilling &
Manufacturing Company
31 E. High St.
East Hampton, CT 06424
(203) 267-4444
FAX: (203) 267-1111

American Colloid
Company
Hwy. 212 West
Belle Fourche, SD 57717
(800) 535 1935
FAX:(605) 892-4880

Aqualon Company
1313 N. Market St.
P.O. Box 8740
Wilmington, DE 19899
(302) 594-6600

-B-

BASF Corp.
100 Cherry Hill Rd.
Parsippany, NJ 07054
(800) 426-8709
FAX: (201) 299 8141

BASF Aktiengesellschaft
D-6700
Ludwigshafen, Germany

Brooks Industries, Inc.
70 Tyler Pl.
S. Plainfield, NJ 07080
(908)561-5200
FAX: (908) 561-9174

-C-

Calgon Corp.
P.O. Box 1346
Pittsburgh, PS 15230
(412) 777-8719
FAX: (412) 777-8104

Croda, Inc.
7 Century Dr.
Parsippany, NJ 07054-4698
(201) 644-4900
FAX: (201) 644-9222

-D-

Dow Chemical Company
2020 Dow Center
Midland, MI 48674
(800) 447-4369

Dow Corning Corp.
P.O. Box 994
Midland, MI 48686-0994
(517) 496-4000
FAX: (517) 496-6487

-F-

FMC Corp.
Pharmaceutical and
Bioscience Division
1735 Market St.
Philadelphia, PA 19103
(215) 299-6534
FAX: (215) 299-6821

-G-

Gattefossé s.a.
36, chemin de Genas
BP 603 69804
Saint-Priest Cedex
France
(33) 72 22 98 00
FAX: (33) 78 90 45 67

Giulini Corp.
105 E. Union Ave
Bound Brook, NJ 08805
(908) 469-6504
FAX: (908) 460-8418

General Electric Company
Silicone Products
260 Hudson River Rd.
Waterford, NY 12188-1926
(518) 233-2264

Givaudan-Roure Corp.
Delawanna Ave.
Clifton, NJ 07015-5034
(201) 365-8000
FAX: (201) 473-1508

BF Goodrich Company
Specialty Polymers &
Chemicals Div.
9911 Brecksville Rd.
Cleveland, OH 44141-3247
(800) 331 1144
FAX: (216) 447-5740

Grain Processing Corp.
Muscatine, IA 52761-1494
(319) 264-4265
FAX (319) 264-4289

-H-

Henkel Corp.
300 Brookside Ave.
Ambler, PA 19002
(800) 531-0815
FAX: (215) 628-1450

Hormel & Company
P.O. Box 800
Austin, MN 55912
(507) 437-5608
FAX: (507 437-5120

Hüls America, Inc.
80 Centennial Ave
Piscataway, NJ 08855-0465
(908) 980-6946
FAX: (908) 980-6970

-I-

ICI Americas, Inc.
Wilmington, DE 19897
(302) 886-3507
FAX: (302) 886-3525

Inolex Chemical Company
Jackson & Swanson Sts.
Philadelphia, PA 19148
(800) 521-9891
FAX: (215) 289-9065

International Specialty
Products
1361 Alps Rd.
Wayne, NJ 07470
(201) 628-4000
FAX: (201) 628-4117

International Sourcing,
Inc.
121 Pleasant Ave.
Upper Saddle River, NJ
07458, (201) 934-8900
FAX: (201) 934 8291

Sutton Laboratories,
ISP Corp.
116 Summit Ave.
P.O. Box 837
Chatham, NJ 07928
(201) 635-1551
FAX: (201) 635-4964

ISP Van Dyk
11 William St.
Belleville, NJ 07109
(201) 450-7722
FAX: (201) 751-2047

-K-

Kelco Division of
Merck & Company, Inc.
8355 Aero Dr.
P.O. Box 23576
San Diego, CA 92123-3576
(619) 292-4900

-L-

Laboratoires
Serobiologiques
161 Chambers Brook Rd.
Somerville, NJ 08876
(908) 218-0330
FAX: (201) 218-0333

Lonza, Inc.
17-17 Route 208
Fairlawn, NJ 07410
(201) 794 2400
FAX: (201) 794-2515

-N-

National Starch &
Chemical Company
10 Finderne Ave.
Bridgewater, NJ 08807
(908) 685-5000
FAX: (908) 685-5005

NIPA Laboratories
3411 Silverside Rd.
Wilmington, DE 19810
(302) 478-1522
FAX: (302) 478-4097

-P-

PPG Industries, Inc.
3938 Porett Dr.
Gurnee, IL 60031
(708) 244-3410
FAX: (708) 249-6344

-R-

Reheis, Inc.
235 Snyder Ave.
Berkeley Heights, NJ
07992
(201) 589-7774
FAX: (201) 589 2615

RHEOX, Inc.
P.O. Box 700
Hightstown, NJ 08520
(609) 443-2000
FAX: (609) 443-2446

Rhône-Poulenc, Inc.
Surfactants & Specialties
CN 7500
Cranbury, NJ 08512-7500
(908) 329-3900
FAX: (908) 320-8165

R.I.T.A Corp.
P.O. Box 585
Woodstock, IL 60098
(815) 337-2500
FAX: (815) 337-2522

Rohm and Haas Company
Independence Mall West
Philadelphia, PA 19105
(215) 592-3000
FAX: (215) 592-3377

-S-

Sandoz Chemicals Corp.
4000 Monroe Rd.
Charlotte, NC 28205
(704) 331-7000
FAX: (704) 377-1064

Southern Clay Products Co.
1212 Church St.
Gonzales, TX 78629
(800) 324-2891
FAX: (210) 672-381

-V-

R.T. Vanderbilt Company, Inc.
30 Winfield St.
Norwalk, CT. 06877
(203) 853-1400
FAX: (203) 853-1452

Index of Manufacturers, Suppliers, Brand Names, and Active Ingredients